Engineering Embedded Systems

Engineering Embedded Systems

Editor

Beniamino Cipriani

scitus academics

Engineering Embedded Systems
Edited by **Beniamino Cipriani**

Printed in 2017

ISBN: 978-1-68117-118-0
Library of Congress Control Number: 2015951991

SCITUS Academics LLC,
616, Corporate Way, Suite 2, 4766,
Valley Cottage, NY 10989

www.scitusacademics.com

Preface

Today's embedded systems development ranges from microprocessor-based control systems, to system-on-chip (SoC) design, and device software development.An embedded system is a computer system with a dedicated function within a larger mechanical or electrical system, often with real-time computing constraints. It is embedded as part of a complete device often including hardware and mechanical parts.

Embedded systems usually are delivered as a combination of software and hardware. Unlike general purpose systems, where the hardware developer does not know which software the system will execute, we are here responsible for both—hardware and software, when developing an embedded system.

The embedded domain offers tremendous job opportunities worldwide. Design of embedded system is an art, and it demands talented people to take up the design challenges keeping the time frames in mind. The biggest challenge faced by the embedded industry today is the lack of skilled manpower in the domain. Though most of our fresh electronics and computer science engineering graduates are highly talented, they lack proper training and knowledge in the embedded domain. Lack of suitable books on the subject is one of the major reasons for this crisis.

This book is concerned with the development of reliable, real-time embedded systems. The particular focus is on the engineering of systems based on time-triggered architectures.The book has been organised in such a way to provide the fundamentals of embedded systems; the steps involved in the design and development of embedded hardware and firmware, and their integration; and the life cycle management of embedded system development.

Table of Contents

CHAPTER 1

An IEEE 1149.X Embedded Test Coprocessor

Ukbagiorgis Iyasu Gebremeskel, José Manuel Martins Ferreira

Faculty of Technology and Maritime Sciences, Buskerud and Vestfold University College, Kongsberg, Norway

ABSTRACT

This paper describes a microprogrammed architecture for an embedded coprocessor that is able to control IEEE 1149.1 to IEEE 1149.7 test infrastructures, and explains how to expand the supported test command set. The coprocessor uses a fast simplex link (FSL) channel to interface a 32-bit MicroBlaze CPU, but it can work with any microprocessor core that accepts this simple FIFO-based interface method. The implementation cost (logic resource usage for a Xilinx Spartan- 6 FPGA) and the performance data (operating frequency) are presented for a test command set comprising two parts: 1) the full IEEE 1149.1 structural test operations; 2) a subset of IEEE 1149.7 operations selected to illustrate the implementation of advanced scan formats.

Table 1 presents test command set that is supported by our embedded test coprocessor. Besides covering all the basic 1149.1 requirements, it includes four additional commands to illustrate 1149.7 test operations. This set offers close equivalents to the main Serial Vector Format (SVF) [10] commands, which are accepted by most test equipment manufacturers. The following section discusses the alternatives available to implement its control path, and offers a detailed description of the micro-programmed architecture that was chosen for this purpose.

A MICRO-PROGRAMMED ARCHITECTURE FOR AN IEEE 1449.7 COPROCESSOR

The first step for implementing a controller architecture supporting the set of commands referred in the previous section consists of building a formal representation of their functionality, bringing into evidence the corresponding control and data flow operations. The data flow operations will determine the blocks required in the coprocessor data path, which will be implemented with regular sequential circuits. There is a higher degree of freedom in what concerns the implementation of the control path, where a hardwired or a microprogrammed architecture can be used to implement the corresponding control flow operations. Notice also that the formal representation of each test command will be influenced by the decision of implementing it in Moore or Mealy form. A Moore machine will only update its outputs (the control signals to the data path) upon the rising edge of the system clock, while Mealy machines can update their outputs at any moment. To illustrate the influence of this choice, Figure 2(a) shows an excerpt of an ASMD chart that illustrates a Moore specification for the SHF1 test command.

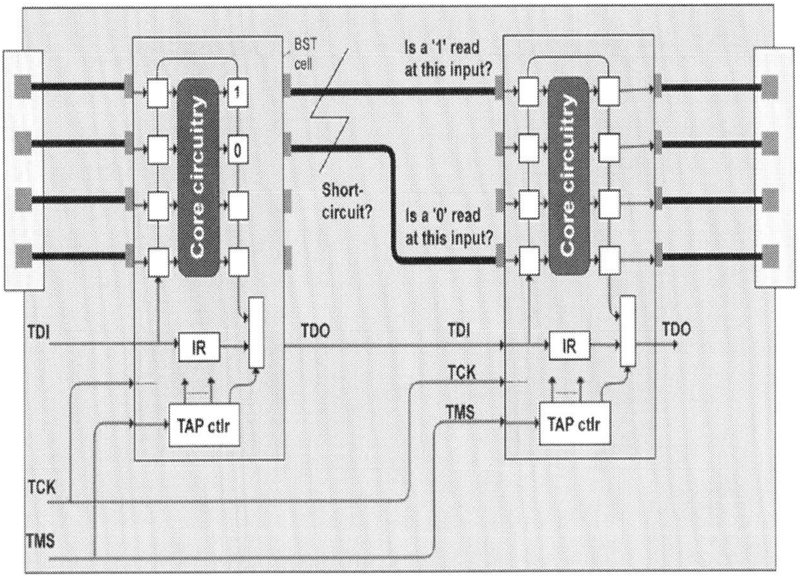

Figure 1. Original IEEE 1149.1 boundary-scan test infrastructure for structural fault detection.

Table 1. IEEE 1149.1/1149.7 test command set supported by the proposed embedded test coprocessor.

Test command	Description
RESET	Takes the boundary-scan logic to Test-Logic-Reset (equivalent to SVF "STATE RESET")
TMS0, TMS1	Set TMS to 0/1 and generate one TCK clock pulse (enabling the implementation of any SVF "STATE" command)
MTCK N	Sets TMS to 0 and generates N TCK pulses (enables the multiple TCK tests carried out by SVF "RUNTEST")
SHF1 N X	Shifts an N-bit bitstream (X) into the [instruction | selected data] register(s) (data bits shifted out are ignored)
SHFCP1 N X, Y, Z	Shifts an N-bit bitstream (X) into the [instruction | selected data] register(s), and compares the output bitstream with its expected response (Y), in all positions indicated by the mask bitstream (Z) (enables the scan & compare "SDR" and "SIR" SVF commands)
ZBS N	Issues N zero bit scan sequences by counting N TMS edge counts (used to generate IEEE 1149.7 commands and to set the control level)
ESC N	Issues N escape sequences (used to select, deselect, reset, and set up other IEEE 1149.7 specific operations)
SHF7O1 N X	Equivalent to SHF1 above, but using the 1149.7 advanced scan format OScan1
SHFCP7O1 N X, Y, Z	Equivalent to SHFCP1 above, but using the 1149.7 advanced scan format OScan1

States 2 and 3 in Figure 2(a) include conditional output boxes used to define Mealy outputs that depend on the condition indicated in the preceding decision box. States 2, 4 and 5 comprise non-empty state boxes used to define Moore outputs, which will be asserted while the controller remains in the corresponding state. The implementation of the same functional behavior in the form of a Moore machine will increase the number of states, since each conditional output box will have to be converted into an equivalent state box, generating a corresponding extra state. The formal representation for the same ASMD chart excerpt, now in the form of a Moore machine, is illustrated in Figure 2(b).

The conversion into a Moore machine increased the number of states by a factor of 2.25 (from 4 to 9). However, this higher number of states does not imply a proportional degradation of cost or performance, be it in the number of logic gates or microprogram memory positions (spatial resources), or in the number of clock cycles (speed). A simple benchmarking of the execution speed can take place by considering the time required to move from state 2 to state 4, assuming that: 1) conditions B and C hold true, and 2) condition D holds false for five times.

The corresponding state transition for the Mealy representation of Figure 2(a) will be 2-3-5-3-5-3-5-3-5-3-5- 3-4, requiring a total of 12 clock cycles. Considering the same initial and final states, and the same assumptions, the equivalent transition for the Moore representation will in this case go through states 2-3-8-5-9-5-9-5-9-5-9-5-9-5-6-7, and requires a total of 15 clock cycles (that is to say, 1.25 times more than the corresponding Mealy representation).

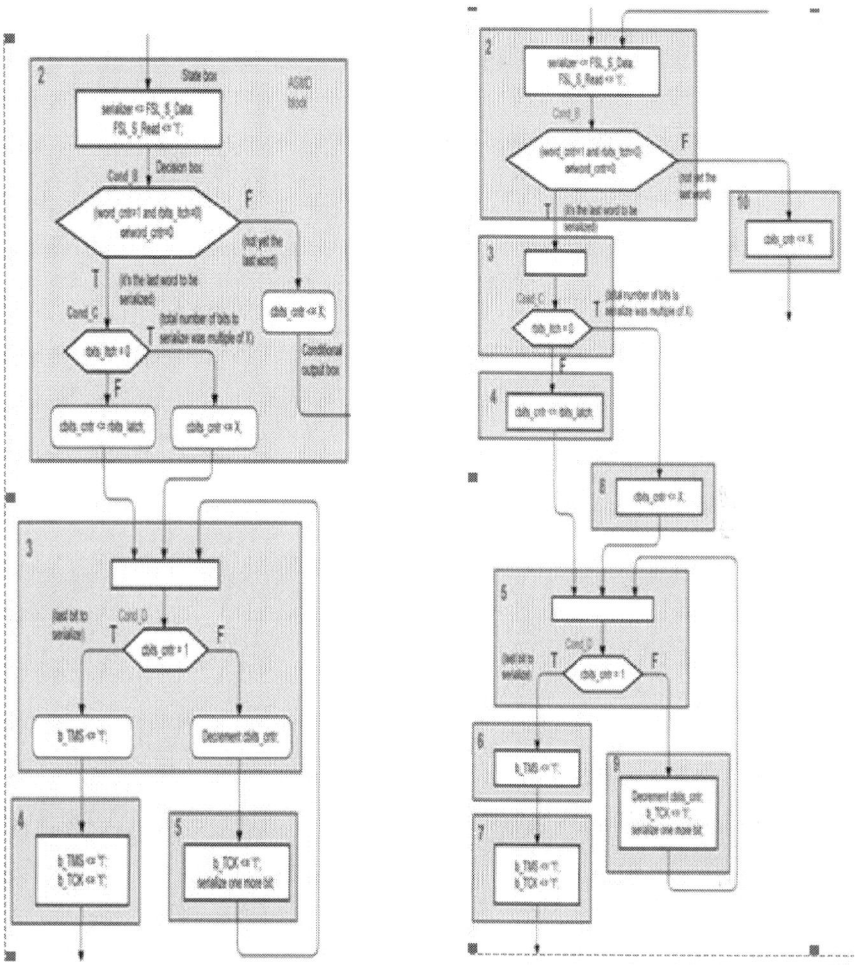

(a) (b)

Figure 2. Partial representation of the ASMD chart for SHF1. (a) Mealy representation. (b) Moore representation.

In order to compare the corresponding implementations in terms of logic resources/FPGA floorspace, a choice will have to be made between a hardwired or a microprogrammed control path architecture. The following reasons explain our choice of a microprogrammed coprocessor architecture:

- A hardwired controller needs to be completely redesigned when a new test command is required, and the designer will have to code each new ASMD chart in the chosen hardware description language (e.g. VHDL, Verilog).
- Even if major differences are not to be expected, a hardwired architecture will correspond to a new sum-ofproducts (or a similar canonical form) structure for each set of test commands, meaning that there will be variations in the critical path and maximum propagation delay (on the contrary, there will be no variation in the control elements of a microprogrammed architecture, since any additional commands correspond solely to further positions added to the microprogram memory).
- Assuming that the "micro-operation set" is able to cover the primitive constructs required to implement new test commands, the designer does not have to write VHDL/Verilog code whenever a new command is added, but simply to translate the respective ASMD chart into the corresponding microprogram memory bank positions.

The basic control path architecture for a microprogrammed implementation is shown in Figure 3(adapted from [11]). In this case, the most significant address bits of the microprogram memory are determined by the test command opcode (which is loaded into the Bank_reg latch), and the ASMD state encoding defines the least significant address bits.

The architecture illustrated in Figure 3 is best in terms of simplicity, and enables a very straightforward implementation of any test command—each state in the ASMD chart will correspond to one word in the microprogram memory, and the operation flow can simply be specified as a sequence of CONTINUE, JUMP or

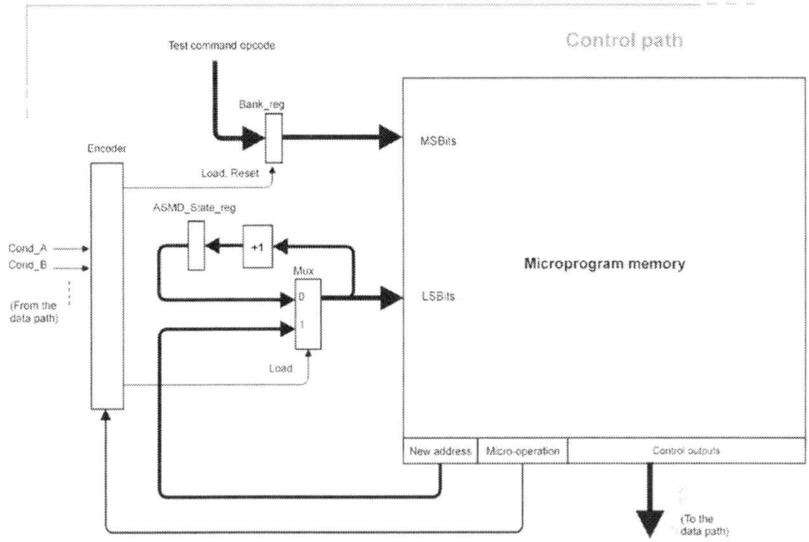

Figure 3. Basic microprogrammed architecture for a Moore control path.

BRANCH IF microinstructions, directing the state transition from beginning to end. However, it can only implement Moore machines, since each ASMD state selects a single microprogram memory position. Since the ASMD blocks frequently contain conditional output boxes, or more than one decision box, this means that the simplicity of the architecture shown in Figure 3 comes at the price of preprocessing the ASMD chart, in order to ensure a pure Moore behavior. The total number of states increases, and so does the number of microprogram memory positions, as well as the number of system clock cycles needed to complete the execution of the corresponding test command (one clock cycle per ASMD chart state).

The general rule for preprocessing the ASMD charts consists of eliminating all conditional output boxes (which will become unconditional outputs specified in a state of their own), and splitting the state when more than one decision box is present. In addition, and since the most significant bits come directly from the test command opcode, the number of microprogram memory positions used to implement each command is fixed. This represents a waste of FPGA floorspace, since the most complex command, with the longest ASMD chart representation, will dictate the number of positions that will be used for all other commands. The control path architecture illustrated inFigure 3 is able to implement any Moore ASMD chart, and would take 9 memory positions to implement the excerpt represented in Figure 2(b). If this was the highest number of states in any command, we would need 4 least significant bits, meaning that the total storage requirements would be given by 2^4*O = 16*O, where O is the number of opcodes comprised in the test command set.

If we want to enable a Mealy implementation, the least significant address bits will have to be driven from data path conditions. This solution enables any number of decision boxes per state, provided that the corresponding conditions are used to drive the least significant address bits of the microprogram memory. The main drawback is that the number of microprogram memory positions will be equal to S*2^D, where S is the number of states, and D is the maximum number of decision boxes existing in a single state (16 memory positions for the example presented earlier). For a test command set with O opcodes, we would have a total microprogram memory storage requirement given by S*2^D*O. While this modification means that we will now have two (or a power of two) microprogram memory positions for each ASMD chart state, the end result is not necessarily an explosion of the microprogram memory space, since state decomposition is restricted to the need of preventing multiple decision boxes per ASMD state (a situation that is rather seldom).

The non-limited simple Mealy architecture enables the fastest implementation. On the other hand, it is the most expensive in terms of microprogram memory storage. The intermediate simple Mealy solution limited to one decision box per state is likewise an intermediate solution in terms of speed vs. microprogram memory storage, while the Moore representation is cheapest in terms of microprogram memory storage. It is also the slowest, although the number of clock cycles is not proportional to the number of states (instead it is dictated by the path through the ASMD chart).Table 2 summarizes the pros and cons of three alternatives: Moore (no conditional outputs allowed, maximum state decomposition), Mealy 1 (one-level Mealy, enabling one decision box and its corresponding conditional output boxes per state), and Mealy 2 (two-level Mealy, enabling up to two decision boxes and four conditional output boxes). Since preprocessing takes place only once, and the difference in speed is small (cf. the simple benchmarking of execution speed that was presented in Figure 2(a) and Figure 2(b)), we adopted the Moore architecture that is represented in Figure 4. Its main advantage over the basic Mealy and basic Moore (Figure 3) architectures consists of eliminating the waste of memory positions that takes place in those two solutions, since the microcode storage requirements are in this case limited to the number of states in the ASMD chart (instead of the being fixed, and dictated by the ASMD chart with the highest number of states).

The specific nature of scan test infrastructures dictates that the number of required primitive constructs (micro-operations) is very small. The data path architecture will therefore have a small number of elements, consisting only of counters, latches and serializers. The conditions associated with the operation of these data path elements are easy to typify, and are limited to detecting if the latches and counters reached one or zero. The proposed test coprocessor architecture may be represented as shown in Figure 5.

Table 2. Pros and cons of Moore, Mealy-1, and Mealy-2 microprogrammed control path implementations.

Topology	Pros	Cons	Comments
Moore	Maximum simplicity	Lowest speed	Preprocessing required
Mealy 1	Good speed upgrade	Higher number of memory positions	Minimum preprocessing
Mealy 2	Marginal speed upgrade over Mealy 1	Highest number of memory positions, most of which will not be used	No preprocessing required (if no. of decision boxes per state is ≤2)

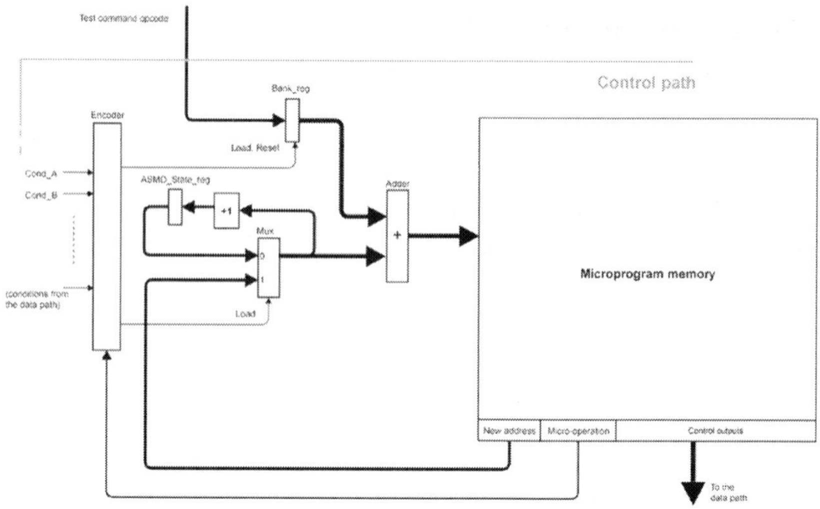

Figure 4. Microprogrammed architecture adopted for the proposed embedded test coprocessor.

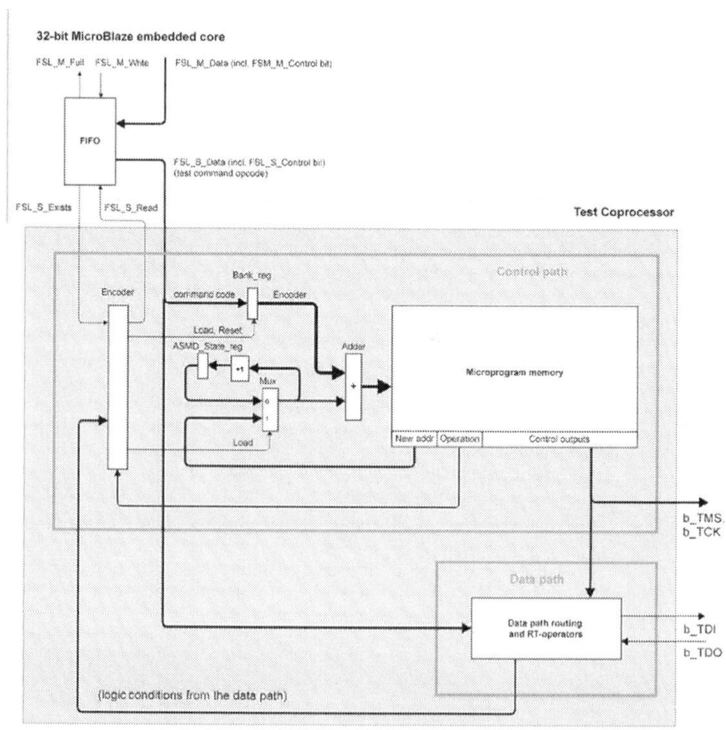

Figure 5. The proposed test coprocessor architecture and interface method.

TEST COMMAND DESIGN

Table 3. Content of the microprogram memory for the MTCK test command.

ASMD state	µprogram ROM	µprogram ROM: control outputs										
	New_Addr	µoperation	iL	iD	rL	cL	cD	cM	sL	sS	bS	bK
0	X	CONT	0	0	0	1	0	0	0	0	0	0
1	Offset to pos. 3	BIF/A	0	0	0	0	0	0	0	0	0	0
2	Offset to END pos.	JUMP	0	0	0	0	0	0	0	0	0	1
3	Offset to pos. 1	JUMP	0	0	0	0	1	0	0	0	0	1

This section presents the micro-operation set that is supported by the microprogrammed control path, and explains the sequence of steps that will enable any designer to add further commands, in order to expand the application domain of the proposed test coprocessor (e.g. further IEEE 1149.7 operations). The microprogrammed control path architecture illustrated in Figure 4 is able to implement the control flow associated to any ASMD chart that is specified in the form of a Moore machine. Each state in the ASMD chart corresponds to one position in the microprogram memory, which comprises the three fields shown in Figure 4 and Figure 5:

- The leftmost bits represent the new address, to be used whenever the next state encoding is different from the current state encoding plus 1 (in which case we'll jump into a new address, instead of incrementing the current address). The number of bits in this field is dictated by the maximum number of states in the ASMD chart of a single test command.
- The middle field contains the micro-operation that represents the required control flow. The implementation of any ASMD chart can be carried out using three basic types of micro-operations: 1) CONTINUE (i.e. increment the current microprogram memory address); 2) JUMP TO ADDRESS (i.e. load the "new address" that is represented in the leftmost bits of the current microprogram memory position); and 3) BRANCH

IF CONDITION TO ADDRESS (jumps to the indicated address if the condition is true, continues to the next address otherwise). This third type actually generates a variety of different micro-operations, i.e. BRANCH IF CONDITION_A TO ADDRESS is formally different from BRANCH IF/CONDITION_A TO ADDRESS (branch if not_condition). Each new test command to be supported will most likely require additional BRANCH IF micro-operations, to cope with the specific conditions associated with its execution. The number of bits in this field is determined by 2 (CONTINUE, JUMP TO ADDRESS) plus the total number of BRANCH IF micro-operations required.

- The rightmost field comprises all the control bits that determine the data flow operations indicated in the ASMD chart. In a horizontal microprogrammed architecture, such as the one that was adopted in this work, the number of bits in this field is equal to the total number of control bits required by the data path elements, plus all additional bits that are directly connected to test access port pins (e.g. board TMS and board TCK).

In order to expand the test command set supported by the coprocessor, the designer shall proceed as follows:

1) Draw the ASMD chart specifying the operation of the required test command.
2) If the ASMD chart contains conditional output boxes (Mealy machine), split those states so as to convert all conditional output boxes into state boxes (convert from Mealy to Moore).
3) Once the Moore ASMD chart is ready, fill in the microprogram memory template with the micro-operations and control bit patterns corresponding to each ASMD state.
4) Update the content of the microprogram memory.

As an illustrative example, Figure 6(a) shows an initial ASMD chart that describes the operation of a MTCK N test command. This command is needed to carry out various types of built-in self-test functions (e.g. all those that rely on pseudo-random pattern generation and parallel signature analysis modes), and generates N test clock (b_TCK) pulses, while keeping b_TMS at "0". MTCK starts by loading the number of required b_TCK pulses (present in the FSL_S_Data bus) into one of the data path counters (cbits_cntr in this case), which will be decremented for each b_TCK pulse. As represented in Figure 6(a), state 1 requires two microprogram memory positions, since the decrement control signal for cbits_cntr may, or may not, be active in this state, according to the condition shown. This example also shows that splitting a state, in order to ensure that a conditional output box is converted into a corresponding state box, does not necessarily increase the number of states in the ASMD chart. Figure 6(b) represents the equivalent Moore ASMD chart, which contains the same number of states. Each state in the chart represented in Figure 6(b)corresponds to a single microprogram memory position. We are now ready to move into the third test command design step, where the microprogram memory template is filled to represent all control and data flow operations associated to the execution of MTCK. Table 3 shows the content of the four microprogram memory positions that are needed to specify the execution of this test command set.

As this example shows, expanding the test command set simply consists of updating the content of the microprogram memory, releasing the designer from the need to understand and modify the VHDL code that describes the control path architecture.

IMPLEMENTATION COST AND PERFORMANCE

To evaluate the overall performance of the proposed embedded test coprocessor, Table 4 and Table 5 show the data that were collected for each test command separately:

- The columns showing data for each test command (the four rightmost columns) correspond to the implementation of a single test command, without the FSL interface.
- Since the FSL interface is predesigned and independent of the proposed microprogrammed architecture, all the tables include two columns showing the implementation data for all test commands when no FSL interface is present, and when a single 1-word 32-bit FIFO interface is added from the MicroBlaze to the test coprocessor.

Table 4 shows the comparison of the logic resources needed by various IEEE 1149.1 test commands with a 4-pin TAP test infrastructure. TMS1 and TMS0 belong to the same group of results, and the same happens with MTCK and RESET, so each second command was omitted in all the tables to improve readability. It is important to notice that the usage of logic resources imposed by the most complex test command (SHFCP1) dictates the cost of the full implementation—the sole implementation of SHFCP1 requires practically the same resources as the full implementation of all the IEEE 1149.1 test commands that are presented in Table1 The number of (iL: iwordL; iD: iwordD; rL: bitsL; cL: cbitsL; cD: cbitsD; cM: cbitsM; sL: serL; sS: serS; bS: b_TMS; bK: b_TCK).

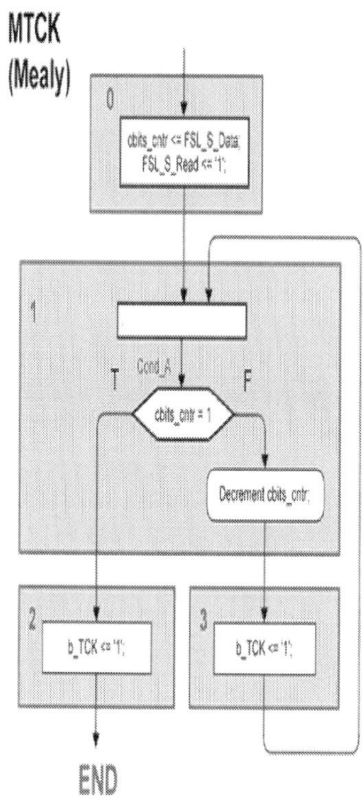

(a)

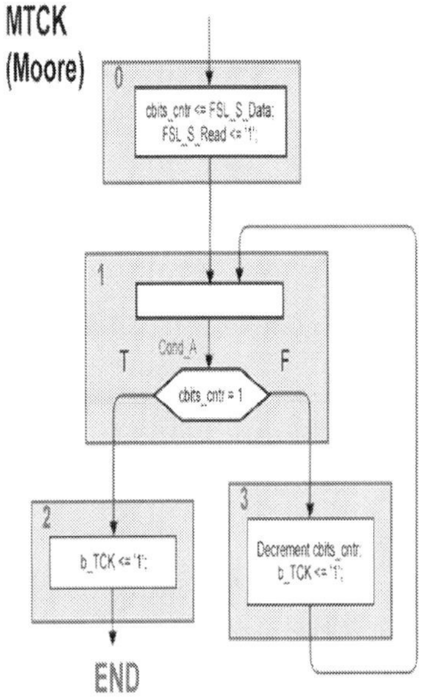

(b)

Figure 6. ASMD chart for MTCK. (a) Mealy representation. (b) Moore representation.

Table 4. Logic resources usage by the IEEE 1149.1 test commands.

Parameter	All (no FSL)	All (one FSL)	Only TMS1	Only MTCK	Only SHF1	Only SHFCP1
No. of slice registers	388 (1%)	374 (2%)	88 -	120 -	274 -	338 -
No. of slice LUTs	383 (4%)	424 (4%)	125 -	207 -	307 -	380 -
No. of memory positions	52	52	4	7	16	23

Table 5. Logic resources usage by the IEEE 1149.7 test commands.

Parameter	All (no FSL)	All (one FSL)	Only ESC	Only ZBS	Only SHF7O1	Only SHFCP7O1
No. of slice registers	160 (1%)	377 (2%)	75 -	75 -	144 -	160 -
No. of slice LUTs	232 (4%)	461 (5%)	179 -	146 -	178 -	224 -
No. of memory positions	98	98	9	23	32	40

microprogram memory positions when the full set of commands is implemented, can be calculated by adding the equivalent value for each command individually, but taking into account that the first two positions are common to all the test commands.

The same comparison has been done for the IEEE 1149.7 test commands, and the summary of the collected data is shown in Table5 This table shows that the usage of logic resources imposed by the most complex test command (SHFCP7O1) again dictates the cost of the full implementation—the sole implementation of SHFCP7O1 requires almost the same resources as the full implementation of all the IEEE 1149.7 test commands that are presented in Table1 The number of microprogram memory positions required when implementing the full set of test commands can be calculated as indicated previously for the 1149.1 commands.

With respect to timing performance, we again notice that the most complex command is the main contributor to determine the minimum period/maximum frequency of the embedded test coprocessor operation. Table 6 shows the timing performance of the implemented IEEE 1149.1 test commands with a 4-pin TAP. The full implementation, in the case of "All (no FSL)", is circa 30% slower than what would correspond to SHFCP1 alone. The minimum period is very close to the maximum combinational path delay, in the case of the sole implementation of the test commands TMS1, TMS0, MTCK, or RESET, but it is less so as we move into the more complex test commands (SHF1 and SHFCP1).

Table 7 shows the timing performance of the implemented IEEE 1149.7 test commands with a 2-pin TAP. One can notice that, as we move to the right, the minimum period/maximum frequency is increasing. This shows that the most complex test command (SHFTCP7O1) is the major contributor to determine the minimum period/ maximum frequency of the operation of the proposed embedded test coprocessor. The data presented in Table 7 shows that the proposed embedded test coprocessor is able to run at TCK frequencies slightly above 70% of the FPGA system clock (which in the case of the Nexys 3™ board runs at 100 MHz).

CONCLUSIONS

Several IEEE 1149.1 test controller solutions have been developed over the years [12] -[17] , but little attention has been given to controller architectures proposed as embedded coprocessors. The work described in this paper offers two main contributions with this respect, by proposing:

1) An embedded test coprocessor that has the capability to test any IEEE 1149.x-compatible system. This coprocessor supports a wide range of testing scenarios for embedded systems based on single or multi-core 32-bit MicroBlaze CPUs, from built-in self-test to online fault detection and diagnosis.
2) A microprogrammed control path architecture for the test coprocessor, enabling a straightforward expansion of the test command set to cope with additional application domains, e.g. those made possible by IEEE 1149.7. The simplicity of the selected architecture enables a relatively fast execution of the test commands, reaching above 70% of the Nexys 3™ board system clock (100 MHz).

Table 6. Timing performance for the IEEE 1149.1 test commands.

Parameter	All (no FSL)	All (one FSL)	Only TMS1	Only MTCK	Only SHF1	Only SHFCP1
Min. period	12.276 ns	12.315 ns	6.053 ns	6.249 ns	7.722 ns	11.246 ns
Max. freq.	81.457 MHz	81.199 MHz	165.211 MHz	160.037 MHz	128.662 MHz	88.917 MHz
Max. comb. path delay	5.157 ns	8.102 ns	5.157 ns	5.385 ns	5.385 ns	5.519 ns

Table 7. Timing performance for the IEEE 1149.7 test commands.

Parameter	All (no FSL)	All (one FSL)	Only ESC	Only ZBS	Only SHF7O1	Only SHFCP7O1
Min. period	12.657 ns	13.655 ns	7.756 ns	8.302 ns	11.469 ns	12.249 ns
Max. freq.	79.005 MHz	73.235 MHz	128.929 MHz	120.449 MHz	87.192 MHz	81.639 MHz
Max. comb. path delay	-	5.94 ns	-	-	-	-

An FSL interface is used to interact with the MicroBlaze for exchanging the command opcode, the arguments of the various functions, and the test result. The FSL is a very popular coprocessor interface method that uses a simple hardware protocol. It is supported by a flexible, yet dedicated instruction set to write to the output and read from the input port. The 32-bit wide FSL bus interface used offers a dedicated point-to-point data streaming interface. Its point-to-point nature and its minimum hardware requirement are the greatest advantages of this interface method. Because the FSL channels are dedicated, no arbitration or bus mastering is required, ensuring an extremely fast interface with very low data latency.

The proposed architecture has the capability to cope with emerging requirements of test and debug standards based on scan test infrastructures. The microprogrammed control path architecture made the embedded test coprocessor scalable, and the FSL interface enabled an optimized solution for IEEE 1149.x test infrastructures, where the most time-consuming functions were executed in hardware, and the remaining functions were implemented by software at a higher abstraction level.

REFERENCES

1. (2001) IEEE Standard Test Access Port and Boundary Scan Architecture. IEEE Std 1149.1-2001, 212 p.http://dx.doi.org/10.1109/IEEESTD.2001.92950
2. (2009) IEEE Standard for Reduced-Pin and Enhanced-Functionality Test Access Port and Boundary-Scan Architecture. IEEE Std 1149.7-2009, 985.http://dx.doi.org/10.1109/IEEESTD.2010.5412866
3. Ley, A.W. (2009) Doing More with Less—An IEEE 1149.7 Embedded Tutorial: Standard for Reduced-Pin and Enhanced-Functionality Test Access Port and Boundary-Scan Architecture. International Test Conference, Austin, 1-6 November 2009, 1-10.http://dx.doi.org/10.1109/TEST.2009.5355572
4. Williams, M. (2009) Low Pin-Count Debug Interfaces for Multi-Device Systems. ARM White Paper. http://goo.gl/KyyVvP
5. Fernandes, F.R., Machado, R.J.S., Ferreira, J.M.M. and Gericota, M.G.O. (2012) IEEE Std 1149.7: What, Why, Where? Proceedings of the 27th Conference on Design of Circuits and Integrated Systems, Avignon, 28-30 November 2012, 118-123.
6. Ungar, L.Y., Bleeker, H., McDermid, J.E. and Hulvershorn, H. (2001) IEEE-1149.x Standards: Achievements vs. Expectations. IEEE Systems Readiness Technology Conference, Valley Forge, 20-23 August 2001, 188-205.http://dx.doi.org/10.1109/AUTEST.2001.948964
7. IEEE Standard for Module Test and Maintenance Bus (MTM-Bus) Protocol (1996) IEEE Std 1149.5-1995. http://dx.doi.org/ 10.1109/IEEESTD.1996.80835
8. IEEE Standard for a Mixed-Signal Test Bus (1999) IEEE Std 1149.4-2010 (Revision of IEEE Std 1149.4-1999).http://dx.doi.org/10.1109/IEEESTD.2011.5738198
9. IEEE Standard for Boundary-Scan Testing of Advanced Digital Networks (2003) IEEE Std 1149.6-2003. http://dx.doi.org/10.1109/IEEESTD.2003.94249
10. Serial Vector Format Specification (1994) Revision E. http://goo.gl/bQQkGI
11. Lynch, M.A. (1993) Microprogrammed State Machine Design. CRC Press, Boca Raton.

12. Yau, C.W. and Jarwala, N. (1990) The Boundary-Scan Master: Target Applications and Functional Requirements. Proceedings of the IEEE International Test Conference, Washington DC, 10-14 September 1990, 311-315.
13. Ferreira, J.M., Matos, J.S. and Pinto, F.S. (1992) Automatic Generation of a Single-Chip Solution for Board-Level BIST of Boundary Scan Boards. Proceedings of the European Design Automation Conference, Brussels, 16-19 March 1992, 154-158.http://dx.doi.org/10.1109/EDAC.1992.205913
14. SN54ACT8990, SN74ACT8990, IEEE STD 1149.1 (JTAG) TAP Masters with 16-Bit Generic Host Interfaces (1997) Texas Instruments Datasheet.
15. Chen, S.J., Zhou, Y., Zhu, D.X. and Guo, S.L. (2011) Design of High-Speed Boundary-Scan Master Controller Base on SOPC. Proceedings of the 2nd International Conference on Mechanic Automation and Control Engineering, Hohhot, 15-17 July 2011, 1195-1198.http://dx.doi.org/10.1109/MACE.2011.5987152.
16. Kaur, M. and Singh, B. (2012) VHDL Implementation of Test Access Port Controller. International Journal of Engineering Science and Technology (IJEST), 4.
17. Cabral, C.J. (2012) Design and Implementation of an IEEE 1149.7-Compliant cJTAG Controller for Debug and Trace Probe. Master of Science in Engineering Report, University of Texas at Austin, Austin, 55 p.

CITATION

Gebremeskel, U. and Ferreira, J. (2014) An IEEE 1149.x Embedded Test Coprocessor. Circuits and Systems, 5, 170-180. doi: 10.4236/cs.2014.57019.

CHAPTER 2

An Architectural Approach with Separation of Concerns to Address Extra-Functional Requirements in the Development of Embedded Real-Time Software Systems

Marco Panunzio, Tullio Vardanega

Department of Mathematics, University of Padova, via Trieste 63, 35121 Padova, Italy

ABSTRACT

A large proportion of the requirements on embedded real-time systems stems from the extra-functional dimensions of time and space determinism, dependability, safety and security, and it is addressed at the software level. The adoption of a sound software architecture provides crucial aid in conveniently apportioning the relevant development concerns. This paper takes a software-centered interpretation of the ISO 42010 notion of architecture, enhancing it with a component model that attributes separate concerns to distinct design views. The component boundary becomes the border between functional and extra-functional concerns. The latter are treated as decorations placed on the outside of components, satisfied by implementation artifacts separate from and composable with the implementation of the component internals. The approach was evaluated by industrial users from several domains, with remarkably positive results.

INTRODUCTION

Embedded real-time systems in general are characterized by two distinctive features: (1) they are resource-constrained, since most often only scarce processing power and memory space are available; and (2) the growing incidence of requirements that address concerns over and above system functionality, in the dimensions of time and space determinism, dependability, safety and, increasingly, security [1].

In this work we collectively refer to those requirements as *extra-functional*, to signify that, while they do not concur to the functional activity of the system, they crucially contribute to the ultimate quality of the system. We also use the term *property* for a feature of an implementation item that provably meets a requirement placed on the specification of that feature. We therefore have extra-functional requirements and extra-functional properties. In this paper we refer to the latter by the shorthand EFP.

Understanding, providing and asserting EFP has an increasingly large and costly effort footprint on the development process in a variety of application domains. The embedded real-time software systems industry therefore seeks ways to accommodate attention for EFP in their otherwise consolidated development practices without breaking the integrity of the overall process.

The central tenet of this work is that the adoption of a sound software architecture helps achieve a clear-cut and composable apportionment of development concerns. With that, EFP can be addressed aside from, yet in coordination with, the functional dimension, in keeping with the established principle of separation of concerns [2].

Interestingly, once the step is taken to put the software architecture at the center of the development strategy, product line concerns can be accommodated by elevating that notion to the stipulation of a *reference* software architecture. We draw from [3] that the reference software architecture is a common and agreed architectural framework capable of addressing all the relevant industrial needs, providing a recurrent solution to the development of a certain class of software systems. To that we add the capability of operating as a single and consistent basis for addressing EFP.

This paper reports on the lessons learned in pursuit of that vision in a large and encompassing research program that progressed along two parallel and complementary lines. One line of the research took place as part of an initiative launched by the European Space Agency (ESA) targeting the definition and realization of a reference software architecture that should steer the development of on-board software for satellites across all of its software supply chain, using the component-based approach described in [4] and [5]. That initiative started from the capture of all product-line needs by the domain stakeholders and concluded with their validation mapping to the chosen reference software architecture. The other line of research occurred within the ARTEMIS JU CHESS project[1] ("Composition with Guarantees for High-integrity Embedded Software Components Assembly" 2009–2012), which aimed at the realization of a model-based component-oriented approach for the development of embedded real-time software systems for telecom, space, and railway applications [6]. All industrial parties from both actions subscribed from the outset to the principle of addressing EFP separately from the functional dimension. In doing so, they were witness to the unifying power of the software architecture concept and to the evidence that EFP were indeed addressed at a distinct level of abstraction as well as at a distinct step in the development process, in overlay to the functional specification of the software.

The remainder of the paper is organized as follows. Section 2 recalls the essential aspects captured in the concept of software architecture and argues why that concept is useful to address concerns in dimensions other than functional. That discussion proceeds into an original interpretation of the notion of reference software architecture and of its founding principles. Section 3 relates the work presented in this paper with state-of-the-art approaches that address similar goals. Section 4 describes the essential details of the proposed approach, with special focus on the way it assists the specification and assures the fulfilment of the extra-functional properties expected of the system. Section 5 presents an instantiation of the reference software architecture to a variety of industrial domains, and discusses how it meets the stakeholders' needs captured as part of the research effort. Section 6 discusses four extensive industrial case studies on which the approach presented in this paper was applied and successively evaluated. Section 7 draws some conclusions and outlines future work.

THE ROLE AND POTENTIAL OF THE SOFTWARE ARCHITECTURE

Common understanding

At odds with its intrinsic centrality, the software engineering practice harbors an exceedingly informal and liberal interpretation of the concept of software architecture. Most practitioners regard software architecture as a synonym to software design: reference [7] collects a large set of community definitions that portray the confusion. In actual fact, the software architecture is a much larger scope than that, with ample bearing on the principles that guide the design and evolution of the software system. The definition given in IEEE 1471, later promoted to ISO 42010 [8] clarifies, to our satisfaction, that an architecture is "the fundamental organization of a system embodied in its components, their relationships to each other, and to the environment, and the principles guiding its design and evolution". When applied to software systems, the definition of software architecture captures well the following concerns:

- *Software decomposition*: the organization of the software in terms of parts, so that every individual part has its own architectural cohesiveness (it addresses a single well-defined part of the problem), and the interactions between parts are minimized so as to reduce unnecessary dependencies (i.e., coupling), hence reducing incidental complexity of understanding, verification and validation, operation and maintenance;
- *Externally visible attributes of software "components"*: those attributes represent features or needs that are specific to

individual components yet can influence other parts of the software or determine properties of the whole. The other attributes, if any, shall remain as (externally invisible) internal details and will not be used for the overall reasoning at the level of the software architecture;

- *Relationship between software "components"*: how components relate to one another in providing services and fulfilling needs;
- *Extra-functional concerns*: the abstraction level at which extra-functional concerns are to be addressed;
- *External interfaces*: the way the software interacts with the external environment (e.g., by commanding sensors or actuators, or serving external interrupts);
- *Principles for the development and evolution of the software*: the software design process that fixes the rules for development, maintenance and evolution, and dictates the supported form of software reuse;
- *The rules in place to warrant the consistent relationship between all of the above concerns*: a methodology capable of encompassing and consistently harmonizing all the aspects listed above; additionally, the methodology shall provide criteria to determine whether a software part can be included in the system, as it conforms to the principles sustained by the architecture, or it shall be rejected, as its inclusion would break system integrity.

A reference software architecture prescribes the concrete form of the software architectures that shape the specific systems for which it was originally developed. The reference software architecture can thus be regarded as a sort of "generic" software architecture that prescribes the founding principles, the underlying methodology and the architectural practices recognized by the domain stakeholders as the baseline solution to the construction of a certain class of software systems in that domain. Symmetrically, the software architecture of one particular system for a given domain can then be regarded as an "instantiation" to the specific system needs of the reference software architecture for that domain.

Narrowing to our context

The reference software architecture chosen for the ESA initiative - and later reflected in the whole span of investigation covered in this work - has a number of distinctive features, which are best introduced here before proceeding further.

First of all, the reference software architecture includes a companion component-oriented development methodology. Software systems in the industrial domains that adopt that methodology are therefore developed by defining (or reusing) components and by creating assemblies of them. The interpretation of components varies with the specific goals of the approach they serve. Numerous definitions for them indeed exist [9],[10] and [11]. Suffice it to say for now that a component is the unit of design and of encapsulation of our approach: we return to this definition later in this paper.

Secondly, the reference software architecture was formulated so as to support the principle of separation of concerns. This is a long-known but much neglected practice first advocated by Dijkstra in [2], which strives to part the various aspects of software design and implementation so as to enable separate reasoning and focused specification for each of them. Separation of concerns is applied to the component model that rests at the very core of the reference software architecture. We consequently pursue it at a level of abstraction much higher than programming, as done for example with aspect-oriented programming [12]. In our approach:

1. Components comprise functional code only, which is to be strictly sequential: extra-functional properties concerning tasking, synchronization and timing (and prospectively fault tolerance) are dealt with outside of the component by the component infrastructure.
2. The extra-functional requirements that the user wishes to set on components are declaratively specified by decoration of component interfaces with an ad hoc annotation language;
3. The realization of extra-functional properties is performed by a code generator, which uses a set of predefined and separately compilable code patterns to generate all of the component infrastructure code for components and their assemblies, and

for the entities that realize extra-functional properties. The implementation of the functional code instead stays under the responsibility of the user.

This extent of separation has two principal benefits: (1) it increases the reuse potential of the software by enabling one and the same functional specification to be reused under different extra-functional requirements; (2) it facilitates the automated generation of vast amounts of complex and delicate infrastructural code addressing concurrency, real-time, communication and component interfacing needs in accord with well-defined styles and fully deterministic rules. Industrial experience shows that benefit (1) is only occasionally achieved, as common and stable specifications, effective component breakdown, clean interface design and consolidation are difficult goals to attain in practice for the functional part of the system, which is intrinsically variable owing to its product-specific nature. Benefit (2) instead becomes available much sooner and at a fraction of the cost of (1), with immediate and tangible benefits.

Finally, the reference software architecture shall help achieve composability and compositionality in a by-construction manner, as opposed to the by-correction style of common practice. With [11] and a little narrowing to the context of this work, we have that: composability is achieved when the properties (needs and obligations) of individual components are preserved on component composition and deployment on the target system; compositionality is achieved when the properties of the system as a whole can be derived (economically and conclusively) as a function of the properties of its constituting components. This work in fact seeks the higher-level goal of composition with guarantees, which with [6] we regard as a form of composability and compositionality assured by static analysis, guaranteed throughout implementation, and actively preserved at run time.

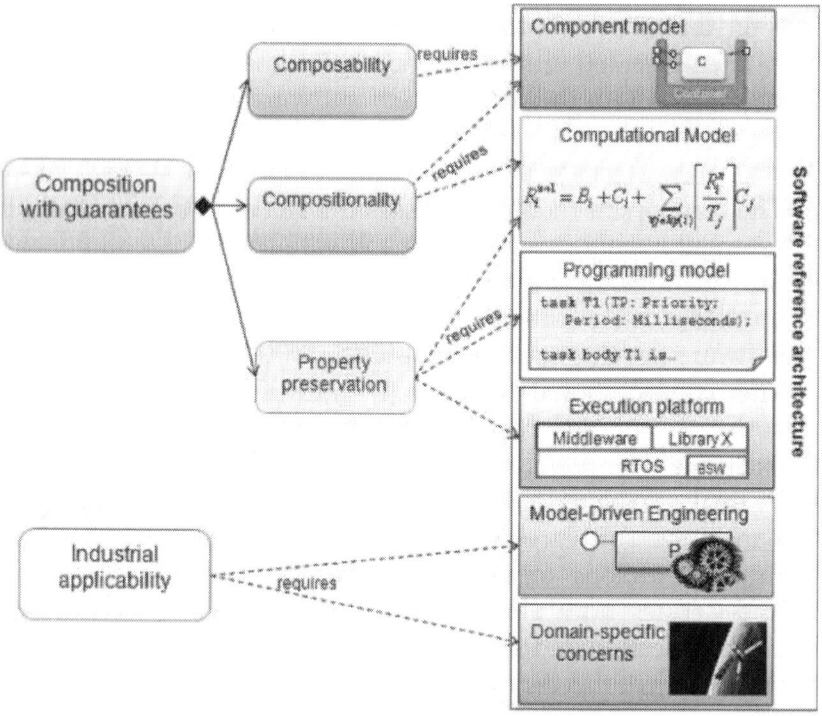

Figure 1. The goals we set to achieve and the constituents of the reference software architecture.

We can therefore contend that, by our choice of essential constituents, we have provided an original interpretation of a reference software architecture that achieves the properties of interest to our industrial stakeholders.

RELATED WORK

There are considerable relations and subtle yet important differences between the premises we have laid down in the preceding sections and the relevant state of the art. The discussion of related work is best broken down in two parts: the state of the art on the combination of component orientation with model-driven engineering, and that which regards how extra-functional properties are addressed in software construction.

Component-oriented model-driven development

We acknowledged earlier in this paper that there is large commonality between our work and the research efforts proceeding from [16]. We consider here those with the closest match.

CoSMIC (Component Synthesis using Model Integrated Computing) [17] addresses the grand challenge of using a modeling and generative programming approach to statically configure and fine tune component middleware tailored to meet quality of service (QoS) requirements for distributed real-time embedded (DRE) applications. As part of that work the authors identify multiple points of integration of model driven techniques with DRE component middleware frameworks, from: (1) configuring and deploying application services end-to-end; (2) composing components into component servers; and (3) configuring application component containers, which fall in the design activity; to synthesizing: (4) application component implementations, (5) dynamic QoS provisioning and adaptation logic, (6) middleware-specific configurations, and (7) middleware implementations, which collectively cover production. Experience shows however that some restrictions and constraints have to be enacted at those points of interaction to achieve "composition with guarantees" at bounded costs, by construction instead of by verification and correction. The cited authors have essentially the same intuition that we placed at the center of our approach, i.e., that model-driven tools can be applied to: (1) analyze relevant characteristics of system behavior, such as scalability, predictability, safety, and security, by using tool-specific model interpreters to feed the analysis tools as needed; (2) synthesize platform-specific code that is customized for particular component middleware and DRE application properties.

PICML (Platform-Independent Component Modeling Language), a central part of CoSMIC [17], is a domain-specific modeling language that enables developers to define component interfaces, QoS parameters and software building rules, and also generates descriptor files that facilitate system deployment. The authors aim to support a component-oriented approach to DRE systems that mitigates the complexity associated with the need to: (1) design consistent component interface definitions; (2) specify valid interactions and connections between components; (3) generate valid component deployment descriptors; (4) ensure that requirements of components are met by target nodes where components are deployed; and (5) guarantee that changes to a system do not leave it in an inconsistent state. PICML has close relations with our work, only we cast the various facets of the endeavor in a single, consistent and coherent software architecture concept, which provides the by-construction guarantees that we seek for our industrial users.

Balasubramanian et al. [18] note that while component middleware technologies raise the level of abstraction used for development, they also promote the decomposition of monolithic systems into assemblies of inter-connected individual components. Functional decomposition of DRE systems into component assemblies and monolithic components helps promote reuse across product lines [19]; however, it can also increase the number of components in the system, with consequences on size and performance. In the cited work the authors propose a "fusion" of analysis and synthesis solutions that help mitigate the problem and obtain a physical assembly, i.e., a component assembly ready for deployment on the physical platform, that has considerably reduced size and improved performance. Once again, this line of work has evident overlaps with ours, with the important difference that the cited works do not seem to pay much attention to achieving a clear-cut apportionment of concerns and responsibilities (and consequently of by-construction guarantees) to distinct elements of the software architecture that should underpin all development.

Addressing extra-functional properties in software construction

Interestingly, what properties are to be seen as composable and what as compositional is not arbitrary, but it rather depends on how a component is defined and on the development methodology devised around that concept.

In recent years, a wealth of approaches studied ways to achieve composability for timing properties. Approaches with composable timing properties typically impose early (and often permanent) constraints on the software architecture. As a consequence, any extra-functional constraint imposed on the functionality provided by a component remains as a permanent characteristic of it and impairs the evolution or the reuse of that component. For example, in the GENESYS project based on a Time-Triggered Architecture [20], a component is a HW/SW subsystem that relates to other components via a "linking interface" (LIF). The LIF shall specify at what time instants communication messages are sent and received by the component, how the messages are ordered, and the rate of message arrival. Concurrency aspects (e.g., tasking) are included in the component specification and are necessary for the correct operation of the component function. Unfortunately, however, they do not emerge at the level of the LIF.

In Integrated Modular Avionics (IMA) - as represented by the ARINC 653 standard [21](which is the predominant incarnation of Time and Space partitioning) - composability for the schedulability results of a task in a partition is achieved at the level of partition, but the scheduling of partitions is rigidly enforced by a cyclic executive, with all the long-known design vulnerabilities of that decision [22].

As a consequence of our choice of seeking separation of concerns, we treat timeliness properties (i.e., response time of operations and end-to-end delay of an activity chain) as compositionality concerns.

The following section expands on this matter and presents our approach to it.

ADDRESSING EXTRA-FUNCTIONAL PROPERTIES

As a distinctive character of this work, EFP are dealt with in accord with the principle of separation of concerns, enacted as follows:

- In the implementation space, with the careful allocation of different concerns to designated software entities: the component, the container and the connector, whose respective role is discussed later in this section;
- In the design space, with the use of design views.

Separation of concerns with different implementation entities

The concepts of component, connector and, less commonly, container, are well-known in the component-oriented community. In relation to them, we adopt a narrower definition that helps serve the principle of separation of concerns [5].

In this work, the component [9] and [10] is the unit of composition. The software system is built by creating an assembly of components, deployed on an execution platform which takes care of their correct execution. In actual fact, the definition by [9] also carries the requirement of "independent deployment of components", but the industrial needs considered in this work do not require it, as also do many other component-oriented approaches.

A component provides a set of functional services and exposes them through a "provided interface". All the services needed by the component from other components or the environment in general are declared in a "required interface". The component is assembled with other components so as to completely satisfy the functional needs of the required interfaces.

Components in this work are pure functional units; they only contain functional code that specifies sequential behavior: time, concurrency, synchronization, distribution and any other extra-functional concerns thus are not included in the component code. All extra-functional concerns are dealt with by the container and the connector, or, via them, by the execution platform; in all cases outside of the component itself.

The container is a software entity that can be regarded as a wrapper around the component and is responsible for the realization of all extra-functional properties that are specified for the component that it embeds. As a consequence, no direct communication to a component can take place, as all communication to them is mediated by the enclosing containers. The container also mediates the access of the component to the executive services it needs from the execution platform.

The connector [23] is the software entity responsible for the interaction between components, which actually is a mediated communication between containers. The role of the connector is to decouple interaction concerns from functional concerns. Components are consequently void of code that deals with interactions with other components. They communicate with endpoints that are set according to the deployment at system elaboration time.

In this way, the functional code of a component can be specified independently of: (1) the components it will be later bound to; (2) the cardinality of the communication; and (3) the location of the other parties. This is necessary as components are designed in isolation and their binding with other components is a later concern, or may vary in different reuse contexts.

The application domains of interest (i.e., space, railways and telecommunication) do not require complex connectors, owing to the static nature of the target systems or of the operational modes that are considered for system modeling. Therefore, support for function/procedure calls, remote message passing through a communication middleware, data access (I/O operations on files in safeguard memory) and event and signals are sufficient to cover virtually all needs. This also means that we do not require an approach for the creation or composition of complex connectors [24]. More complex connector kinds are nevertheless necessary when stronger guarantees on remote communication are required (e.g., by introducing domain-specific communication protocols), for location and representation transparency in more heterogeneous systems.

Fig. 2 and Fig. 3 outline the envisaged development process.

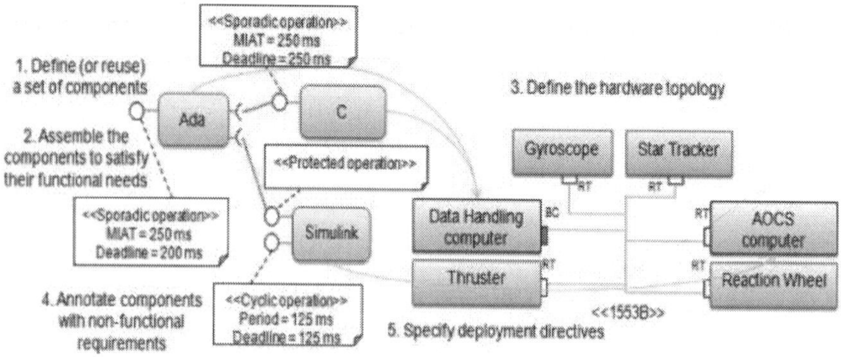

Figure 2. A sketch of the actions that the user performs in the design space in the proposed approach.

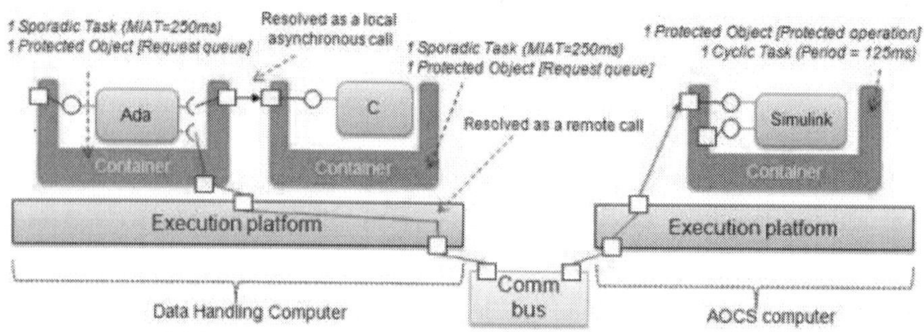

Figure 3. . An outline of the automated generation of implementation entities (containers and connectors).

The designer first creates or reuses a set of components, and assembles them to fulfill their functional needs. In parallel, the designer defines the hardware and the hardware topology of the system. Then they attach to component interfaces a set of extra-functional annotations.

We describe in detail the specification of extra-functional in the next section, as it is of fundamental importance to our approach. Finally, the designer specifies a set of deployment directives to allocate components to computational nodes. At that stage, the software systems description (the model) can be submitted to various extra-functional analysis; those currently of interest to our industrial stakeholders include: schedulability analysis, state-based dependability analysis; Failure Mode and Effect Analysis (FMEA) and Failure Mode, Effects and Criticality Analysis (FMECA).

EFP are only declared on components. An example of declaration syntax, for the concerns of tasking, concurrency and real-time is presented in [5]. The semantics of those declarations is informed by the chosen computational model, in our case the Ravenscar Computational Model (RCM) [25]. The RCM postulates a concurrent mode of execution that restricts the classic sporadic task model of real-time scheduling theory to a form directly amenable to static analysis in the time and space dimension and efficiently implementable by a small-footprint real-time kernel. By adopting the RCM, the system resulting from the assembly of components can be statically analyzed for its end-to-end response times, including the overhead of containers and connectors, with response-time analysis [26].

EFP are realised by model-to-text transformations that generate containers and connectors according to a deterministic (hence provable) mapping of the extra-functional annotations set on components to code implementation artifacts [27].

Using the component bindings and the deployment information, the same pass of model-to-text transformation generates the connectors that realize the desired communication semantics between components, possibly relying on the service stack of the execution platform for transparent remote communication.

Fig. 2 depicts examples of EFP annotations. Owing to the model-to-text transformation rules in place (see [27] for a full description of them), the two sporadic operations declared on the two components implemented in Ada and in C shown in the example, require each the creation of a container which comprises: (1) a protected object (a queue protected with the Immediate Ceiling Protocol [28]) used to deposit incoming requests of execution; (2) a sporadic task released at every 250 ms. (so as to respect the minimum separation between subsequent executions, called Minimum Inter-arrival Time, MIAT for short), to fetch the first request in the queue and execute it, or to block on an empty queue. The operation marked as "protected" on the component implemented in Simulink, requires a protected object (or in alternative form, a semaphore) to provide execution of it in mutual exclusion. The protected object is generated in the container, which also hosts a cyclic task to periodically execute the operation marked as "cyclic". Fig. 3 depicts the containers produced from applying model-to-text transformation to the example.

Remarkably, the generated containers are invariant with respect to chosen deployment of components: it is connectors that take care of understanding it. In our example, the communication between the component implemented in Ada and the one implemented in C is resolved as a local asynchronous call (local, as the components are deployed on the same node; asynchronous, as the provided operation is sporadic, and therefore has its own thread of execution). The call from the Ada component to the Simulink component is mapped instead to a remote communication with message passing, as the destination component is deployed on a remote node.

The Ravenscar Computational Model restricts the nature and behavior of low-level implementation entities such as tasks, monitors, semaphores, underneath components. By using annotations on higher-abstraction entities such as components and interfaces, which are bound to the RCM semantics via model transformations, we are able to faithfully create lower-level analysis models on which accurate model-based schedulability analysis can be performed [29], while gaining the advantages of reasoning at the abstraction level of components in the user model.

Design views and specification of extra-functional concerns

According to ISO 42010 [8], the architectural description "is organized into one or more constituents called views", each of them being "a representation of a whole system from the perspective of a related set of concerns". Continuing to quote, each view is the expression of a specific viewpoint, that is "a specification of the conventions for constructing and using a view. A pattern or template from which to develop individual views by establishing the purposes and audience for a view and the techniques for its creation and analysis".

Our work rests on these concepts, tailoring them to component-oriented development supported by Model-Driven Engineering tool environments.

In current approaches such as state-of-the-art SysML [30] technology, views and viewpoints can be defined by the user, and consequently vary for distinct systems of the same class. This contrasts with the choice made in this work - arguably more consistent with the cited reference - whereby views are statically defined by the developers of the methodology, in agreement with the domain stakeholders.

If during the construction of the development approach, each view can be ratified as the expression of a single concern, then views become effective and powerful means to enforce separation of concerns in the specification of the system.

Design views also are the means to enforce a given flow of activities for software development: a development process envisions multiple design phases to be carried out - some in potential parallelism, others in some specific ordering - quite possibly under the responsibility of distinct development actors. If the passage from one design phase to the next crosses the boundary of responsibility between different actors with distinct concerns, then design views come as a very handy means to enforce this boundary.

For example, it may be prescribed that the specification of certain design entities or attributes in one view cannot be made without the prior definition of other entities in other views. The corresponding methodological framework should explicitly recognize this prescription and provide means to specify the desired design views accordingly. In MDE terms, the corresponding design views would be implemented by specifying visualization and modification rights for the metaclasses that refer to the entities in question, as defined by the metamodel for the chosen design language. Those rights would then be enforced in the various views to any concrete instance of that metaclass in the design model under development.

Technology solutions exist to evaluate constraints at model creation time, but they do not suffice to satisfy our needs. Not only in fact we require the capability of evaluating constraints related to entities in the model and their static relationships, but we also need to maintain a "stateful context" of when and by whom a model modification (e.g., creation of a new element, specification of an attribute) is performed: we capture those needs in the active design view. The present state of practice for MDE technology does not provide a complete solution to those needs: any current provision for design views is unable to work at the level of the metamodel - where this capability is best provided - and consequently resorts to augmenting the model editor in manners that are neither standard nor portable.

In this work the following design views were defined:

- *Data view*, for the description of data types and events (which are messages generated or received by a component following a publish/subscribe model);
- *Component view*, for the definition of interfaces, components and the bindings between components to fulfill their functional needs;
- *Hardware view*, for the specification of hardware and the network topology;
- *Deployment view*, for the specification of the allocation of components to nodes;
- *Extra-functional view*, where extra-functional annotations are attached to the functional description of components. This view may be further divided in sub-views in order to keep a cohesive control on the concerns of interest: in the reported work, the real-time view and the dependability view were defined as specific sub-views in the extra-functional dimension;
- *Dependability view*, a (sub-) view where specific extra-functional annotations enable various forms of model-based dependability analysis (i.e., error modeling, state-based analysis, FMEA, FMECA) [31] [32]. This view was implemented for industrial users in the CHESS project.
- *Behavioral view*, an optional view where the designer can specify in the model the functional code of component implementations using UML state machines and generate C++ code for it [33]. This view was implemented for industrial users in the CHESS project: in the absence of this view the designer can associate

Ada or C/C++ source code written manually to a component implementation;

- *Space-specific view*, an optional view where the designer can specify the use of services related to spacecraft commandability and observability as defined in space-specific standards (e.g., [34]). This view was required for use by the European Space Agency and its industrial suppliers;
- *Railways-specific view*, an optional view where specific extra-functional decorations make it possible to address a number of railways-specific connection protocol concerns.

No telecom-specific view was deemed necessary in CHESS. The technical requirements of the domain were either already addressed by the component model or – when not addressed yet – carried sufficient interest for other industrial domains to be promoted to the domain-neutral part of the component model. An interesting example of a need promoted to core concern was for the decoration for "multiplicities" with component instances and ports. This was handy syntactic sugar to lessen the cluttering of the design space and the burden of specification for the user.

Another reason for not needing a telecom-specific view in CHESS reflects the specific case study chosen by the industrial partner (see Section 5), which solely exercised the domain-neutral part of the component model in conjunction with a reference execution platform for the domain.

All in all, the experience gained from the industrial case studies presented in Section 6suggests that addressing domain-specific protocols at the component level is best done via dedicated, domain-specific views: their presence facilitates the specialized implementation of the relevant needs in connectors and the associated component infrastructure, and segregates their specification away from the core domain-neutral part of the component model.

Fig. 4 recapitulates the design views of our approach, and highlights the precedence constraints between them (i.e., the entities to be defined in a view depend on information from another view).

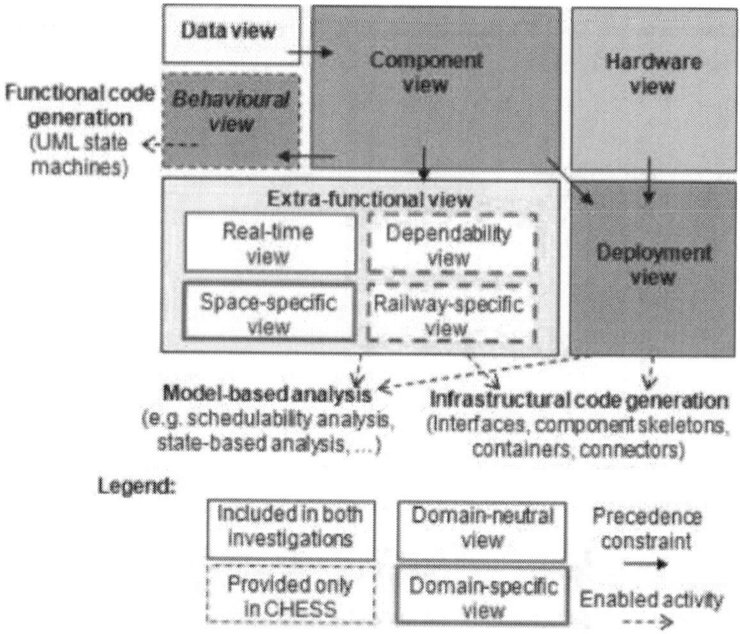

Figure 4. . The design views provided by our approach.

Some considerations are in order at this point.

Firstly, the segregation of different concerns to different design views proved to be an effective means to enact a determined design flow across design activities.

Secondly, all domain-specific views were born in relation to extra-functional concerns. This is interesting in two ways: (1) it shows the goodness of fit of the core domain-neutral part of the component model, which was effectively shared by three different application domains; (2) it also shows that ease of activation of a domain-specific view by simply adding domain-specific annotations to the total catalog of extra-functional annotations.

Thirdly, separation of concerns as interpreted in this work enables the generation of functional code completely separate from the rest of the component infrastructural code (i.e., interfaces, component interfaces, containers, connectors). Because of this separation, functional code generation (or implementation) can be performed independently of any later deployment and extra-functional consideration. Moreover, the implementation of components can be directly submitted to functional verification and validation (V&V) (which encompasses unit testing and integration testing of component assemblies) as dictated by the applicable domain standards. The functional V&V results stay valid even if later the extra-functional annotations of components change, as the functional code does not suffer any impact. Only a few functional tests must be performed only after deployment information are specified: those related to the numerical precision of functional algorithms, as they necessarily require execution on target.

INSTANTIATION TO MULTIPLE INDUSTRIAL DOMAINS

The vision and the technology described so far in this paper were submitted to the evaluation of industrial users from three distinct domains: space (by two distinct research actions, in the contexts of the European Space Agency and of the CHESS projects), telecom and railways; the latter two also from CHESS. It was therefore necessary to submit the notion of reference software architecture and its constituents presented in Section 2.3 to the industrial users to learn from them which would stay common and which should require adaptations and changes.

Component Model. The same component model, defined in [4] and [5], was retained in all industrial domains, though with two different implementations.

The ESA side of the work created an incarnation of the component model centered on a domain-specific metamodel (nicknamed "SCM", short for Space Component Model) and developed an associated graphical editor based on the Obeo Designer framework.[2]Obeo Designer supports the static definition of viewpoints, which covered a sizeable part of the needs for design views. It was possible to specify per-view diagrams, graphical representation for design entities, automations and constraints.

In CHESS instead, industrial users required the use of open standards, which resulted in the adoption of UML and the MARTE profile (augmented with project-specific stereotypes) and in the use of the Papyrus[3] technology. CHESS-specific plug-ins were developed so as to provide automation capabilities, model validation checks, and support for the design views specified in the previous section.

Computational Model. All industrial users, except for telecom, converged on the adoption of the Ravenscar Computational Model (RCM) [25]. The two space case studies and the one for railways therefore explored the use of model-based schedulability analysis [29]. The support they were offered allowed the results of the analysis to be seamlessly back propagated to the design model in the form of decoration attributes of the design entities, which could be inspected without needing access to external tools.

Programming Model. The two space case studies and the one for railways choose the Ravenscar Profile as programming model, which was rather natural after choosing RCM for a computational model. A set of RCM-conformant and property-preserving code archetypes programmed in Ada meet the requirements of the programming model[35] and [27]. They are used for the automated generation of the component infrastructural software that comprises: component skeletons, component interfaces, containers (including tasking and synchronization code) and connectors for communication concerns, by which we provide for enforcement of the extra-functional properties specified in the design model communication.

Execution platform. This is where the differences appeared among industrial domains.

In the European Space Agency case study, the execution platform comprised: (i) an implementation of the Packet Utilization Standard (PUS) services [34], which, in contrast to traditional code-centric approaches [36], are configured in the PUS view supported in this work, thus at the level of abstraction of the user model; (ii) the Spacecraft Onboard Interface Services[4] (SOIS), a set of standard services used for domain-specific message transmission on network/subnetworks, and communication with distribution transparency; (iii) an open-source operating system conforming with the RCM, for providing tasking, communication and synchronization primitives.

In the CHESS space case study, the target execution platform included: (i) a proprietary implementation of the PUS services; (ii) a proprietary middleware for distribution transparency and communication; (iii) a proprietary operating system with execution semantics conforming with the RCM.

The railways case study used a representative open-source middleware for distributed communication and Ada-level Ravenscar concurrency for execution on Linux.

Not conforming choices. The telecom case study made different choices for all aspects other than the component model and yet stayed broadly consistent with the notion of reference software architecture promoted in this work. The main interest of the industrial user was to try the component model for architectural modeling, the behavioral view for functional specification via state machines, and direct C++ code generation to OSE, a telecom-specific target operating system. [5] That choice covered the concerns addressed in the reference software architecture by the computational model and the programming model, which however were not of high interest of that industrial user.

Fig. 5 recapitulates the choices of each case study.

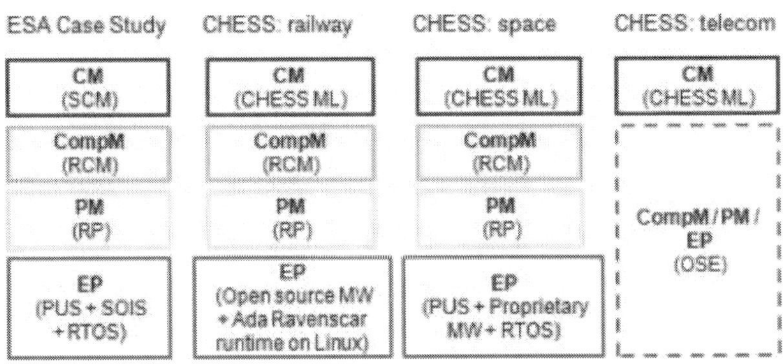

Figure 5. . What the case studies picked as the four constituents of the reference software architecture.

CASE STUDIES AND INDUSTRIAL EVALUATION

Not all the case studies were end-to-end (from design to code generation). Their common goal was to validate the core approach, the common component model and its goodness of fit for the reference industrial process of interest. Additionally, each industrial user focused on aspects of the methodology of their specific interest.

The evaluation focused on assessing process-related results and product-related results; the industrial teams were also asked to evaluate the proposed approach according to some miscellaneous criteria.

Process-related results

- *PR-1*: Adoption of design views in the design process;
- *PR-2*: Adoption of a component-oriented process with a set of rigorous design steps;
- *PR-3*: Early schedulability analysis with back-propagation of results;
- *PR-4*: Complete generation of extra-functional code related to tasking, synchronization and time-related aspects;
- *PR-5*: Potential for model-based functional code generation;
- •*PR-6*: Use of domain-specific views for separate specification of domain-specific concerns.

Product-related results

- *PD-1*: Use of containers for realization of extra-functional properties separately from component code;
- *PD-2*: Use of connectors for implementing communication needs and domain-specific properties;

Miscellaneous criteria

- *M-1*: Maturity of the methodology
- *M-2*: Increased productivity
- *M-3*: Learning curve

European Space Agency case study

This case study re-engineered a subset of the on-board software of a small yet representative Earth Observation satellite, operating in Low Earth Orbit and including an optical payload to capture images on demand.

The involved industrial user focused on evaluating whether: (1) the component model is able to express the needs for the development of on-board software for satellites; (2) the component model is able to accommodate space-specific needs (namely, the PUS services) in a manner that is consistent with the rest of the approach, and to provide the designer with specification means at the right level of abstraction in the space-specific view.

The case study was performed over 6 calendar months by a senior engineer without previous experience on model-driven, component-oriented methodologies of a small-size on-board software prime contractor, with support from a part-time consultant. The conclusions of the investigation were reviewed by two large software and system prime contractors of the space domain.

The experience provided some interesting feedback.

Firstly, fitting an existing, code-centered design baseline into the novel approach may not be straightforward: the strict adherence to separation of concerns at code level promoted with our method may entail considerable re-engineering for a code base that couples aspects that our approach views as orthogonal and separately addressed (i.e., functional code vs. tasking and synchronization code). Emancipation towards separation of concerns also requires careful attention in determining the right level of abstraction for expressing components in the new software architectural design: all too often, code-centered mentality tends to use components as tasks, which is a patent inversion of abstraction.

Secondly, an important amount of architectural reflection is needed to achieve a specification of components of the most convenient "size" (in quantity of provided services). Designing a great number of "small" components, although individually more easily amenable to reuse, may introduce a lot of complexity in the design

The case study focused on modeling two subsystems: (a) an "analyzer", which performs the analysis of the GSM-R signal; (b) a "receiver", which receives the analyzed data on signal quality on board and performs the appropriate actions in response.

The "analyzer" is deployed on a laptop, the "receiver" on a dedicated board (simulated with a laptop in the case study).

The case study was small-sized, yet centered on several key features: (i) support for the creation of components written either in C or Ada; (ii) support for multi-node systems; (iii) support for a railways-specific communication protocol to regulate communication from (a) to (b).

Also this case study was based on the re-engineering of an existing code base, and was performed by a software engineer without previous experience in component-oriented methods, plus support from a lead engineer on dependability-related modeling. The case study was performed in two iterations for a total duration of 4 months.

Summary and evaluation

Table 1 recapitulates the key aspects of the proposed methodology and how they were evaluated by the feedback obtained from the four industrial case studies.

Table 1. I (Demonstrated and considered interesting and promising); Ad (Demonstrated and considered Adequate); NA (Not applicable); HM (High-maturity); m (moderate); l (low); H (high learning curve); M (moderate learning curve).

ID	Result/criteria short name	ESA case study	CHESS: telecom	CHESS: railways	CHESS : space
PR-1	Adoption of design views	I	I		I
PR-2	Component-oriented process	Ad	Ad	Ad	I
PR-3	Model-based schedulability analysis				I
PR-4	Automated generation of extra-functional code			I	I
PR-5	Model-based functional code generation		I		
PR-6	Use of domain-specific views	I	NA	I	I
PD-1	Containers for separate realization of extra-functional properties	I		I	I
PD-2	Connectors for communication needs and domain-specific properties			I	
M-1	Maturity of methodology	HM	HM	HM	HM
M-2	Increased productivity	m	m	l	l
M-3	Learning curve	H	H	H	M

Process-related results

PR-1 is cited as an important factor in the telecom and space domain feedback of the CHESS project and in the ESA investigation.

PR-2 is cited by all case studies as adequate or interesting for the realization of their target system. It helps to split the intellectual work in manageable parts that can be realized (or further refined) independently by the designers.

PR-3 was successfully exercised in the space case study of CHESS, which highlighted its potential if it were introduced in the stakeholder's industrial process.

PR-4 was successfully achieved in the railways case study of CHESS. The code generation was also evaluated by the CHESS space case study, and deemed satisfactorily and promising.

PR-5 was demonstrated successfully on the CHESS telecom case study, where C++ code was generated for functional state machines associated to component implementations ([33]). The approach explored in the cited publication is very promising indeed, though it currently lacks integration with the techniques in place for the generation of extra-functional code (for containers, connectors, and component infrastructure). The difficulty with filling that gap is purely technical, with no bearing on fundamentals, contingent on the stabilization of the heterogeneous technologies used for the various steps of required transformations. Achieving this integration is part of our future work in a recently started follow-on project to CHESS.

PR-6 was realized for the space and railways domains. The three involved case studies confirmed the usefulness of the concept, which was considered interesting for specifying in a coherent manner their domain-specific needs directly in the user model, where the software is being specified.

Product-related results

PD-1 was demonstrated in the CHESS space and railways case studies and in the ESA investigation, and considered a key result.

PD-2 was demonstrated by implementing a basic communication protocol for the railways use case. The protocol can be parametrized by activating the railways view and applying railways-specific annotations to component bindings. Notably, no code of components and containers needed modification, as the only changes when the protocol is activated are carefully isolated in the code generated for the involved connectors. Furthermore, in the ESA case study, we realized a sizeable set of the PUS services, and demonstrated that it is possible to control and parametrize them directly at the abstraction level of the design space.

Miscellaneous evaluation criteria

For what concerns M-1, the maturity of the methodology is considered high by the feedback of all case studies.

For what concerns M-2, the productivity gains are still considered low to moderate. This feedback is mostly due to the prototypal nature of the two toolsets and to the lack of collaborative features for the modeling activity. We however are satisfied by this fair evaluation, as the feedback by industrial users highlights how this is exclusively a technological problem (and in fact the methodology itself was considered mature).

Finally, the learning curve of the approach (M-3) is considered from moderate to high. This highly depends on the previous exposure of the project partner to MDE and component-oriented approaches (the CHESS space partner is quite familiar with those paradigms, the space partner of the ESA investigation and the telecom and railways partners of CHESS were not). We maintain that most of the difficulty stems from the raised abstraction level proper of any model-driven engineering approach, which can be hard to master for an engineer exposed to it for the first time or after using UML and its derivatives only for documentation.

Furthermore, the difficulty of fitting a novel development approach into the pre-existent industrial process of a stakeholder - an aspect often ignored or downplayed in the scientific literature - shall not be underestimated. In particular, experimenting the novel approach in the place of a "traditional" approach (i.e., not already based on model-driven engineering and without separation of concerns) by re-engineering an existing code base. instead of starting a new architectural design from scratch, bears much hidden risk: the mindset of the actors stays more oriented to "twisting and bending" the novel approach to fit the existing design and code, rather than to applying it to ripe its full advantages.

This feedback also highlighted that an increasing effort shall be devoted to training and dissemination activities, especially in the form of tutorials or reference guides (for example, with the description of architectural patterns for solving recurrent design problems using the component model and the reference software architecture).

CONCLUSIONS AND FUTURE WORK

This paper presented the results of two closely interconnected research efforts targeting embedded real-time systems: one effort was set to establish a reference software architecture for the on-board software of future satellite missions procured by the European Space Agency; the other defined a component-oriented methodology fit for use in the development of telecommunication, space, and railways software applications.

The industrial users in both efforts concurred from the start to the idea of separating out the treatment of extra-functional concerns from the rest of development issues. What was needed to that end was a systematic way to guarantee that the resulting products would meet the requirements by construction as opposed to by correction. The proposed solution centered on the formulation and the adoption of a reference software architecture based on four distinct yet closely related constituents: a component model, a computational model, a programming model and a execution platform. Specific instances were then proposed for each of those architectural constituents.

Separation of concerns was enacted at two levels of development.

At the design level, by segregating distinct concerns to separate design views, where design views are a manifest asset of the adopted component model. Each design view enforces specific creation, visualization and modification rights on design entities. Those rights are attributed according to the concern that the entity in question is an expression of. An entity relevant to a concern will allow creation and modification rights on itself in the design view attached to that concern: entities not relevant of that concern will instead be just read-only or even possibly plainly invisible in that particular design view. Extra-functional properties are defined as annotations attached to the outside of the component specification or to its exposed interfaces. The component boundary thus becomes the border between functional and extra-functional concerns.

At the implementation level, the realization of extra-functional concerns is delegated to two other software entities that pertain to the real-time architecture of the component framework: the container (which addresses tasking, synchronization and time-related concerns among others); and the connector (which addresses communication and interaction concerns). In the proposed approach, containers and connectors are automatically generated via model transformations, in a manner that conforms by construction to the programming model that implements the computational model chosen in the reference software architecture, and that preserves the desired execution semantics when on the target platform.

The industrial users involved in the research effort adopted a single component model and used all of its domain-neutral core features in their respective case studies. They were able to address a selection of their domain-specific needs via the addition of domain-specific views to the component model. This capability provided convincing evidence of the successful enactment of separation of concerns and of its ability to produce fully composable parts, all individually conforming to common design principles, and all consistently and verifiably contributing to end-to-end performance.

In the proposed approach the handling of extra-functional concerns stays strictly under the control by the design environment and it is performed via automated model transformation. This provision ensures the consistency of the approach and in particular - because of the congruent adoption of a computational model, a programming model and a conforming execution platform - warrants consistency between the software system as analyzed and the system implementation at run time.

The authors are now working on an extension of the component model that supports hierarchical components, with particular attention to ensuring that the specification and realization of extra-functional properties can work with top-down decomposition as well as bottom-up composition. Future plans also include further enhancement to the domain-specific extension capabilities of the component model.

ACKNOWLEDGMENTS

This work was supported by the Networking/Partnering Initiative of the European Space Agency and by the CHESS project under ARTEMIS JU grant No. 216682

REFERENCES

1. I. Crnkovic, M. Larsson, O. Preiss, Concerning predictability in dependable component-based systems: classification of quality attributes, in: Architecting Dependable Systems III, Vol. 3549 of Lecture Notes in Computer Science, Springer, Berlin Heidelberg, 2004, pp. 257–278.
2. E.W. Dijkstra, On the role of scientific thought, in: E.W. Dijkstra (Ed.), Selected Writings on Computing: A Personal Perspective, Springer-Verlag, New York Inc, 1982, ISBN 0-387-90652-5, pp. 60–66.
3. P. Kruchten, The Rational Unified Process: An Introduction, third ed., AddisonWesley Longman Publishing Co. Inc., 2003, ISBN 0-321-19770-4.
4. M. Panunzio, T. Vardanega, A component model for on-board software applications, in: Proceedings of the 36th Euromicro Conference on Software Engineering and Advanced Applications, IEEE Computer Society, 2010, pp. 57–64.
5. M. Panunzio, Definition, Realization and Evaluation of Software Reference Architecture for Use in Space Application, Ph.D. Thesis, University of Bologna, Italy, 2011,
6. T. Vardanega, Property preservation and composition with guarantees: from ASSERT to CHESS, in: Proc. of the 12th IEEE International Symposium on Object/Component/Service-Oriented Real-Time Distributed Computing, IEEE Computer Society, 2009, pp. 125–132.
7. Software Engineering Institute (Ed.), Defining Software Architecture – Modern, Classic, and Bibliographic Definitions, SEI – Carnegie Mellon
8. ISO/IEC, Systems and software engineering – Architecture description, ISO/IEC 42010:2011 (2011).

9. C. Szyperski, Component Software: Beyond Object-Oriented Programming, 2nd ed., Addison-Wesley Professional, Boston, 2002, ISBN 0-201-74572-0.
10. M. Chaudron, I. Crnkovic, Component-based software engineering, in: H. van Vliet (Ed.), Software Engineering: Principles and Practice, Wiley, 2008, ISBN 0- 470-03146-8 (chapter 18).
11. J. Sifakis, A framework for component-based construction extended abstract, in: Proceedings of the 3rd IEEE International Conference on Software Engineering and Formal Methods, IEEE Computer Society, 2005, ISBN 0- 7695-2435-4, pp. 293–300.
12. G. Kiczales, J. Lamping, A. Mendhekar, C. Maeda, C. Lopes, J.-M. Loingtier, J. Irwin, Aspect-oriented programming, in: Proceedings of the 11th European Conference on Object-Oriented Programming, Vol. 1241 of Lecture Notes in Computer Science, Springer, Berlin Heidelberg, 1997, ISBN 978-3-540-63089- 0, pp. 220–242.
13. M. Panunzio, T. Vardanega, On component-based development and highintegrity real-time systems, in: Proceedings of the 15th International Conference on Embedded and Real-Time Computing Systems and Applications, IEEE Computer Society, 2009, ISBN 978-0-7695-3787-0, pp. 79–84.
14. T. Vardanega, Development of on-board embedded real-time systems: an engineering approach, Technical Report ESA STR-260, European Space Agency, ISBN 90-9092-334-2, 1999.
15. M. Bordin, T. Vardanega, Correctness by construction for high-integrity realtime systems: a metamodel-driven approach, in: Proceedings of the 12th International Conference on Reliable Software Technologies - Ada-Europe, Vol. 4498 of Lecture Notes in Computer Science, Springer, Berlin Heidelberg, 2007, ISBN 978-3-540-73229-7, pp. 114–127.
16. D.C. Schmidt, Model-driven engineering, IEEE Comput. 39 (2) (2006) 25–31.
17. K. Balasubramanian, J. Balasubramanian, J. Parsons, A.S. Gokhale, D.C. Schmidt, A platform-independent component modeling language for distributed realtime and embedded systems, in: Proceedings of the 11th IEEE Real-Time and Embedded Technology and Applications Symposium, IEEE Computer Society, 2005, ISBN 0-7695-2302-1, pp. 190–199.
18. K. Balasubramanian, D.C. Schmidt, Physical assembly mapper: a model-driven optimization tool for QoS-enabled component middleware, in: Proc. of the 14th IEEE Real-Time and Embedded Technology and Applications Symposium, IEEE Computer Society, 2008, ISBN 978-0-7695-3146-5, pp. 123–134.
19. R. van Ommering, Building product populations with software components, in: Proceedings of the 24th International Conference on

Software Engineering, ACM Press, 2002, ISBN 1-58113-472-X, pp. 255–265.

20. H. Kopetz, N. Suri, Compositional design of RT systems: a conceptual basis for specification of linking interfaces, in: Proceedings of the 6th IEEE International Symposium on Object-Oriented Real-Time Distributed Computing, IEEE Computer Society, 2003, ISBN 0-7695-1928-8, pp. 51–60.
21. Aeronautical Radio, Incorporated, ARINC Specification 653-1: Avionics Application Software Standard Interface (2003).
22. G. Douglass, Locke, software architecture for hard real-time applications: cyclic executives vs. fixed priority executives, Real-Time Syst. 4 (1) (1992) 37–53.
23. N.R. Mehta, N. Medvidovic, S. Phadke, Towards a taxonomy of software connectors, in: Proceedings of the 22nd International Conference on Software Engineering, ACM, 2000, ISBN 1-58113-206-9, pp. 178–187.
24. B. Spitznagel, D. Garlan, A compositional approach for constructing connectors, in: Working IEEE/IFIP Conference on Software Architecture, IEEE Computer Society, 2001, ISBN 0-7695-1360-3, pp. 148–157.
25. A. Burns, B. Dobbing, T. Vardanega, Guide for the use of the Ada Ravenscar profile in high integrity systems, Technical Report YCS-2003-348, University of York, 2003.
26. J.C. Palencia, M. González, Harbour, schedulability analysis for tasks with static and dynamic offsets, in: Proceedings of the 19th IEEE Real-Time Systems Symposium, IEEE Computer Society, 1998, ISBN 0-7803-5243-2, pp. 26–37.
27. M. Panunzio, T. Vardanega, Ada Ravenscar code archetypes for componentoriented development, in: Proceedings of the 17th International Conference on Reliable Software Technologies - Ada-Europe, Vol. 7308 of Lecture Notes in Computer Science, Springer, 2012, ISBN 978-3-642-30597-9, pp. 1–17.
28. J.B. Goodenough, L. Sha, The priority ceiling protocol: a method for minimizing the blocking of high priority Ada tasks, in: Proceedings of the 2nd International Workshop on Real-time Ada Issues, ACM, 1988, ISBN 0-89791-295-0, pp. 20–31.
29. M. Bordin, M. Panunzio, T. Vardanega, Fitting schedulability analysis theory into model-driven engineering, in: Proceedings of the 20th Euromicro Conference on Real-Time Systems, IEEE Computer Society, 2008, ISBN 978-0- 7695-3298-1, pp. 135–144.
30. Object Management Group, SysML specification - version 1.3,

31. L. Montecchi, P. Lollini, A. Bondavalli, Dependability concerns in model-driven engineering, in: Proceedings of the 14th IEEE International Symposium on Object/Component/Service-Oriented Real-Time Distributed Computing Workshops, IEEE Computer Society, 2011, ISBN 978-0-7695-4377-2, pp. 254–263.
32. L. Montecchi, P. Lollini, A. Bondavalli, Towards a MDE transformation workflow for dependability analysis, in: Proceedings of 16th IEEE International Conference on the Engineering of Complex Computer Systems, IEEE Computer Society, 2011, ISBN 978-0-7695-4381-9, pp. 157–166.
33. F. Ciccozzi, A. Cicchetti, M. Krekola, M. Sjödin, Generation of correct-byconstruction code from design models for embedded systems, in: Proceedings of the 6th IEEE International Symposium on Industrial Embedded Systems, IEEE, 2011, ISBN 978-1-61284-818-1, pp. 63–66.
34. European Cooperation for Space Standardization (ECSS), Space Engineering - Ground systems and operations - Telemetry and telecommand packet utilization, ECSS-E-70-41A, 2003.
35. E. Mezzetti, M. Panunzio, T. Vardanega, Preservation of timing properties with the Ada Ravenscar profile, in: Proceedings of the 15th International Conference on Reliable Software Technologies - Ada-Europe, Vol. 6106 of Lecture Notes in Computer Science, 2010, pp. 153–166, ISBN 978-3-642-13549-1.
36. T. Vardanega, G. Caspersen, Engineering software reuse for on-board embedded real-time systems, Softw. - Pract. Exp. 32 (3) (2002) 233-264. ISSN 0038-064.

CITATION

Marco Panunzio, Tullio Vardanega, An architectural approach with separation of concerns to address extra-functional requirements in the development of embedded real-time software systems, Journal of Systems Architecture, Volume 60, Issue 9, October 2014, Pages 770-781, ISSN 1383-7621, http://dx.doi.org/10.1016/j.sysarc.2014.06.001.

CHAPTER 3

A Component-Based Process with Separation of Concerns for the Development of Embedded Real-Time Software Systems

Marco Panunzio, Tullio Vardanega

Department of Mathematics, University of Padova, via Trieste 63, 35121 Padova, Italy

Highlights

- We propose a component-based approach for embedded real-time software systems.
- The approach meets requirements from the space, railway and telecom domains.
- The approach enforces separation of concerns throughout the development process.
- The approach supports model-based analysis and code generation.
- The approach was assessed in four case studies in two parallel research projects.

ABSTRACT

Numerous component models have been proposed in the literature, a testimony of a subject domain rich with technical and scientific challenges, and considerable potential. Unfortunately however, the reported level of adoption has been comparatively low. Where successes were had, they were largely facilitated by the manifest endorsement, where not the mandate, by relevant stakeholders, either internal to the industrial adopter or with authority over the application domain. The work presented in this paper stems from a comprehensive initiative taken by the European Space Agency (ESA) and its industrial suppliers. This initiative also enjoyed significant synergy with interests shown for similar goals by the telecommunications and railways domain, thanks to the interaction between two parallel project frameworks. The ESA effort aimed at favouring the adoption of a software reference architecture across its software supply chain. The center of that strategy revolves around a component model and the software development process that builds on it. This paper presents the rationale, the design and implementation choices made in their conception, as well as the feedback obtained from a number of industrial case studies that assessed them.

INTRODUCTION

Non-functional concerns such as time and space predictability, dependability, safety, and more recently security, have an increasingly large incidence on system development in high-integrity application domains such as avionics and space, railways, telecom and, prospectively, automotive. Several of those needs are addressed by software.

This trait places stringent requirements at process and product level, corresponded by onerous verification and validation (V&V) needs. Industry therefore seeks ways to contain the cost of development while strengthening the guarantees on the result.

Among the various solutions proposed to that end, the adoption of Model-Driven Engineering (MDE) Schmidt (2006) has fared rather well by measure of interest and success. Evidence collected in domain-specific initiatives (cf. e.g., (Bordin and Vardanega, 2007, Panunzio and Vardanega, 2007 and Bordin et al., 2008)) shows that the higher level of abstraction in the design process facilitated by MDE allows addressing non-functional concerns earlier in the development, thereby enabling proactive analysis, maturation and consolidation of the software design. Moreover, the automation capabilities of the MDE infrastructure may ease the generation of lower-level design artifacts and enable the automated generation of source code products of certain quality.

In the space arena specifically, experience gained in the ASSERT[1] project persuaded the European Space Agency (ESA) and its main system and software suppliers that for the adoption of MDE methods to produce tangible benefits, a software reference architecture common to all development stakeholders should be established first.

ISO/IEC/(IEEE) (2007) defines an architecture as composed of: (a) the fundamental organization of a system embodied in its components; (b) their relationships to each other, and to the environment; and (c) the principles guiding its design and evolution. On that basis, Panunzio and Vardanega (2013) regards the concept of software reference architecture as proceeding from: (i) a component model, to design the software as a composition of individually verifiable and reusable software units; (ii) a computational model, to relate the design entities of the component model, their non-functional needs for concurrency, time and space, to a framework of analysis techniques which assures that the architectural description is statically analysable in the dimensions of interest by construction; (iii) a programming model, to ensure that the implementation of the design entities obeys the semantics, the assumptions and the constraints of the computational model; (iv) a conforming execution platform, which actively preserves at run time the system and software properties asserted by static analysis and it is able to notify and react to possible violations of them.

ESA and their industrial partners decided to explore how well that concept could serve as a basis for their MDE-adoption initiative, and saw their effort complemented by the parallel launch of the CHESS project.[2] Those two efforts successfully collaborated in the

To this end, we restrict the expressive power of the user up front by propagating to the design space the constraints that emanate from the proposed development approach and by enforcing them actively, so that the resulting model is correct by construction in the dimensions of interest.

Separation of concerns

A long-known but much neglected practice first advocated by Dijkstra in Dijkstra (1982), separation of concerns strives to separate different aspects of software design and implementation to enable separate reasoning and focused specification for each of them. We apply that notion to our component model, by making the following distinctive choices:

1. components comprise functional (sequential) code only: the non-functional needs with bearing on run-time behaviour (such as e.g., tasking, synchronization, timing) are dealt with outside of the component by the component infrastructure, expressed in terms of containers, connectors and their runtime support;
2. the non-functional requirements that the user wishes to set on components are declaratively specified by decoration of component interfaces with a specific annotation language, which currently addresses concurrency, synchronization, time and memory, and, to a lesser extent via the work of other authors, fault tolerance and safety concerns; non-functional requirements that specialised model-based analysis ascertains that can be met by the component infrastructure on the execution platform of choice are elevated to non-functional properties;
3. using predefined and separately compilable code templates, a code generator that operates in the back-end of the component model builds all of the component infrastructure that embeds the user components, their assemblies, and the component services that help satisfy the non-functional properties asserted by analysis.

The extent of separation of concerns that ensues from these choices has two principal benefits: (1) it increases the reuse potential of the software by enabling one and the same functional specification (corresponding to one or more components) to be reused under different non-functional requirements (corresponding to instantiations of component infrastructure); (2) it facilitates the automated generation of vast amounts of complex and delicate infrastructural code addressing non-functional concerns with bearing on run-time behaviour, which in this paper are limited to concurrency, real-time, communication and component interfacing needs, in accord with well-defined styles and fully deterministic rules, with obvious benefits in terms of important life-cycle properties including readability, traceability, maintainability.

Experience shows that benefit (1) is much more difficult to achieve than it may seem, as it requires substantial effort to arrive at common and stable specifications, effective component breakdown, clean interface design and consolidation for the functional part of the system. Conversely, benefit (2) becomes available much sooner and at a fraction of the cost of (1), with immediate and tangible benefits. This is a crucial point to our whole concept.

Composition

Our definition of software reference architecture builds on the premise that general-purpose programming languages are not and cannot be component models themselves. In addition to missing some fundamental capabilities in fact - most notably multi-concern interface semantics for the specification of provided and required interfaces - programming languages operate at a level of abstraction that is lower (i.e., too implementation specific) to that proper of component models. Conversely, restricted and specialized profiles of apt programming languages may help develop specific parts of the correct-by-construction code artifacts that implement component-based systems.

Our approach aims at achieving the properties of composability and compositionality. When composability and compositionality can be assured by static analysis, guaranteed throughout implementation, and actively preserved at run time, we may speak of*composition with guarantees* (Vardanega, 2009), which is our grander goal here.

In accord with Sifakis (2005), we maintain that composability is achieved when the designated properties of individual components, captured in terms of needs and obligations, are preserved on component composition, deployment on target and execution. Our components operate in the functional space only. They therefore express strictly sequential semantics. The signature of their methods determines how the invocation occurs functionally. Components are stateful and their state is comprised within the component. From the interaction perspective, components are black boxes that only expose provided and required interfaces. These provisions help warrant functional composability, void of non-functional semantics, hence without concurrency, interleaving, synchronisation (among other concerns with bearing on run-time behaviour). Functional composability, a narrower view of composability tout-court, warrants that the properties held by the sequential execution of the functional interfaces provided by individual components verified in isolation are preserved, in the functional dimension, when components are composed by the binding of their matching required and provided interfaces. Yet there is more to composability than just functional concerns.

We address the non-functional dimension, again limited to concerns with bearing on run-time behaviour, in two steps. The specification of non-functional behavior is super-imposed on component interfaces in a manner that preserves their original functional semantics and enriches it with non-functional semantics separately realized by the container that encapsulates the component. Indeed, it is the interface provider (as opposed to the caller) that determines the semantics of the invocation, including for the effects that the execution has on the component state. This prescription is crucial to ensuring that interface decoration *adds* to the functional semantics expressed by the component itself, instead of possibly conflicting with it. Interface decoration is a conveyor of semantic enrichment which takes effect in safeguarding the non-functional behavior of components and their functional binding to one another. The syntax used for interface decoration is comparatively arbitrary. Yet the semantics that decorations capture must match the execution semantics stipulated by the computational model to which the component model is attached.

The computational models that fit our needs (1) help extend composability to the non-functional dimensions of interest, with concurrency and real-time especially considered in this paper, and (2) make it possible to take a compositional view of how execution occurs at system level. In accord with Sifakis (2005), we regard compositionality to be achieved when the properties of the system as a whole can be determined as a function of the properties of the constituting components. The binding of a computational model to the component model allows the execution semantics of components with added non-functional descriptors to be fully understood in the face of concurrency, interleaving, contention, synchronisation (and of any other dimension covered by the computational model).

From the real-time perspective, example properties that may be attached to the provided interfaces of a component, include worst-case execution time (WCET), period, deadline. Looking at each of these three properties helps appreciate what composability and compositionality signify in this context. The WCET is a local property of the program (that is, the service attached to the interface in question): composability in the time dimension (Puschner et al., 2009) is achieved if the interfering effect caused by the presence of other components in the system does not prevent a safe and tight WCET bound to be determined for every single interface service. Period requirements are composable so long as the execution platform can sustain them without incurring unacceptable jitter. This property can be asserted statically by analysis of features of the execution platform. For a scheduling algorithm that allows interleaving - which is what we assume in this work - the satisfaction of deadline requirements is a property of execution that can only be asserted compositionally. For the scheduling algorithm adopted by the execution platform of choice, and reflected in the concurrency and real-time semantics of the decoration attributes that emanate from the corresponding computational model, schedulability analysis compositionally determines how the completion time of individual services, effected by the interference effects caused by job-level interleaving, relates to the applicable deadline. This analysis is intrinsically compositional as it uses a decomposition of the system that allows singling out the local properties of interest and uses them to determine the effect that they will have globally.

By its very nature, the computational model considers entities that belong to the implementation level (e.g., tasks, protected objects, semaphores). In the design of a component model therefore, and especially for its use in an MDE process, a higher-level representation of those entities must be provided that: (i) does not pollute the user model with entities that pertain to a lower level of abstraction; (ii) meaningfully represents those entities and their semantics; (iii) ensures that it always is possible to correctly transform the information set by the designer in the higher-level representation into entities recognized by the computational model. In our approach, needs (i) and (ii) are addressed by concentrating the representation of the required semantics in interface decoration attributes. Needs (iii) instead is addressed as part of the generation of correct-by-construction code artifacts.

Software entities

The real-time architecture of our component model features three distinct software entities: the component, the container and the connector.

Components and connectors are present in most component-oriented approaches: a wealth of literature discusses their various possible flavours (see for example (Szyperski, 2002, Lau and Wang, 2007 and Mehta et al., 2000)). Containers have a much lesser prominence in the literature, perhaps a token of the insufficient penetration of the concept of separation of concerns in component-based software engineering. They are used in approaches (like the one presented herein) which present an endogenous treatment of non-functional properties (i.e., outside of the component) (Crnkovic et al., 2011). Industrial-level examples of containers exist, for example OSGI containers and Enterprise JavaBeans (EJB) containers (although those are more intended as a run-time environment for components rather than a per-component wrapper).

In our context, the component addresses exclusively functional and algorithmic concerns, the connector is used to address interactions concerns, and the container is responsible for the realization of the non-functional concerns, with regard to concurrency (tasking and synchronization), real-time and reconfiguration aspects.

The component is the only entity that appears at design level. Containers and connectors pertain to the implementation level and get attached to components by way of fully automated transformations. This attachment allows components to interact among themselves and with the execution platform once deployed in the run-time environment.

Component

Chaudron and Crnkovic (2008) defines a software component as "a software building block that conforms to a component model". The authors of the cited work maintain that: "a Component Model defines standards for (i) properties that individual components must satisfy and (ii) methods, and possibly mechanisms, for composing components".

To us, the component is the unit of composition throughout the system development process, from conception, reuse, refinement and aggregation, which all pertain to the modeling phase, to system building, which for us corresponds to producing the full sources of all component implementations, and of their connectors and containers, ready for compilation and linking. The software system is built as an assembly of components, deployed on an execution platform which takes care of their execution needs. The requirement of "independent deployment of components" entailed by the definition bySzyperski (2002) is currently not a core requirement for us (and neither is in many other component-oriented approaches). This stance matches the practices in use at the target industrial domains of interest, which all require verified and validated static builds of the system, which treat upgrades and reconfigurations outside of the development phase, in the long-lasting operation and maintenance phase.

A component provides a set of functional services and exposes them through a "provided interface"; the services needed from other components or the environment in general are declared in a "required interface". The component is assembled with other components so as to satisfy the functional needs of its required interfaces. Components can also use event-based communication with a publish-subscribe communication paradigm. Components can register to an "event service" in order to receive notifications of events emitted by other components.

Non-functional attribute descriptors are added to component interfaces to specify the non-functional properties desired for the execution of the corresponding services. Those attributes are taken from a fixed language of declarative specifications. The semantics of that declarative language emanates from the chosen computational model: the Ravenscar Computational Model (RCM) (Burns et al., 2014) in our case. That provision has three important consequences:

1) it fully informs the model transformations that automatically produce the containers and connectors which serve to realize the non-functional requirements set on the component interfaces and the binding among them;
2) it enables the execution of schedulability analysis directly on the model of components; and
3) it warrants full consistency between the specification of non-functional concerns and their realization in the implementation.

The left part of Fig. 1 depicts two components with their contract interfaces and a component binding between them.

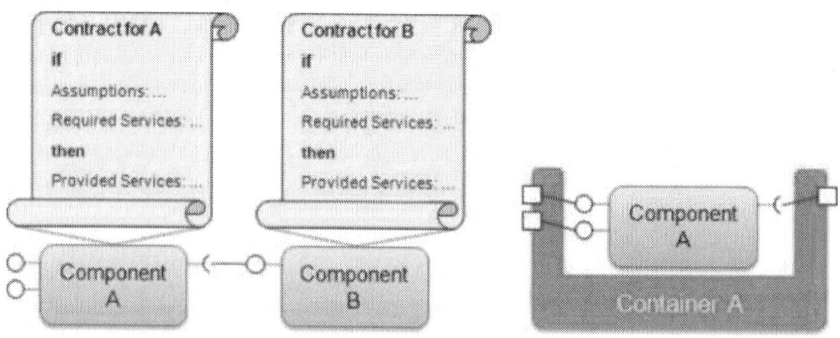

Figure 1. Two components with their interfaces and component binding and a container.

Container

The container (see Fig. 1) is a software entity that can be regarded as a wrapper around the component, which is directly responsible for the realization of the non-functional properties specified on the component that it embeds. In programming terms, the relation between the container and the component is the same as that determined by inversion of control,[3] the style of software construction where reusable code (the container) controls the execution of problem-specific code (the component).

The container exposes the same provided and required interfaces as the enclosed component, through interface "promotion" and "subsumption" relations from the services of the component to the equivalent services in the container. With interface promotion, the container is able to prefix the component's execution with what it takes to realise the non-functional semantics attached to the relevant component interface. With interface subsumption, the container is able to intercept the interface calls made by the component and transparently forward them to the container that wraps the target component. As a result of that provision, there is no direct communication to a component, since all communication between them is mediated by the enclosing containers. The container also mediates the access of the component to the executive services it needs from the execution platform.

Binding between components, as we said already, is statically defined at model level (no dynamic loading of components is allowed), yet – in line with the *inversion of control*principle – it takes effect at software initialization time, when initialization of containers and their external connections takes effect.

The right part of Fig. 1 depicts a container and its embedded component and also shows the promotion and subsumption of the corresponding interfaces.

Connector

The connector (Mehta et al., 2000) is the software entity responsible for the interaction between components, which actually is a mediated communication between containers. Connectors allow separating interaction concerns from functional concerns. We maintain that this separation is beneficial in that the user only needs to *specify* the interaction style and semantics to be established in the binding between components, without having to produce the code-level – or model - for it, thus with benefits in terms of correctness guarantees and assured performance. Components are consequently void of code that deals with interactions with other components.

A connector decouples the component from the other endpoints of the communication. In this way, the functional code of a component can be specified independently of: (1) the components it will be eventually bound to; (2) the cardinality of the communication; and (3) the location of the other parties. This is necessary as components are designed in isolation and their binding with other components is a later concern, or may vary in different reuse contexts.

The nature of our target systems reduces the variety of necessary connectors to a few basic kinds, which are required to perform function/procedure calls, remote message passing or data access (I/O operations on files in safeguard memory). This also means that we do not require an approach for the creation or composition of complex connectors (Spitznagel and Garlan, 2001). More complex connector kinds are necessary when communication between components implemented in different languages is carried out, guarantees on remote communication are required and for location and representation transparency in more heterogeneous systems.

Fig. 2 depicts a connector that regulates the interaction between two containers. The figure also shows that there can never be direct connection of a component with the execution platform, as the connection is always mediated by the container.

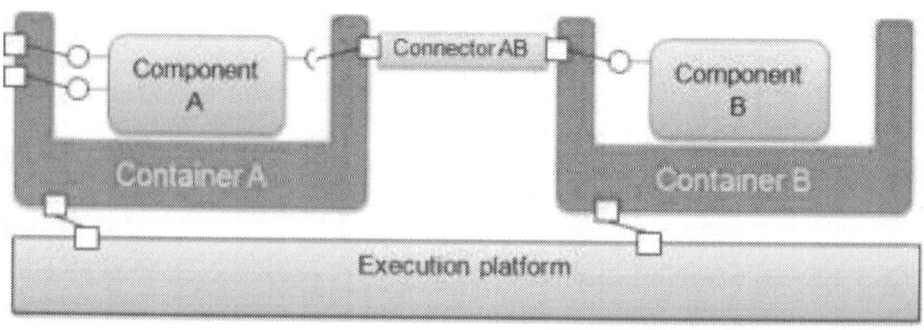

Figure 2. A connector that realizes the communication between two containers on behalf of their respective components. The figure shows also the underlying execution platform.

Execution platform

With the generic term "execution platform" we identify the middleware, the real-time operating system/kernel (RTOS/RTK), communication drivers and the board support package (BSP) for a given hardware platform. For the purposes of this paper, we consider the execution platform as a single monolithic block, and we just categorize the services it is to provide to our design and implementation entities.

The concerns addressed by a given platform service determine the software entity entitled to use it. We in fact classify platform services according to their user:

- services for containers: they are used by the containers to enforce or monitor non-functional properties. For example: tasking primitives, synchronization primitives, time-related primitives, timers.
- services for connectors: they are the implementation of communication means, constructs to transparently handle physical distribution across processing units, libraries for translation of data encoding.
- services for components: they are (infrastructural) services intended for the functional service of components; typifying examples of such services include: access to the system time for time-stamping data; context management and data recovery. To use those services, components do not access the execution platform directly: access to them is mediated by the corresponding container.

In contrast to components, whose implementation only includes sequential code and thus is independent of the execution platform (no direct calls to OS primitives or other execution platform services are allowed), the implementation of containers and connectors necessarily depends on the target platform, to which it allows components to statically bind. It is therefore necessary to create specific implementation of containers and connectors for each execution platform of interest.

Component models

The primary purpose of the component model in the work presented in this paper is to favour reuse with guarantees, in the connotation given in Section 2.1, while embracing a model-based development paradigm. In terms of the rich classification proposed inCrnkovic et al. (2011), our component model (i) addresses the modeling and implementation phases of the development life cycle, (ii) is independent from the programming language, (iii) provides constructs for interface specification, (iv) allows expressing a limited set of interaction patterns, (v) supports specification, composition and analysis of non-functional properties, and (vi) is special-purpose as intended to high-integrity embedded real-time systems.

At the cost of some redundancy with the review provided in the cited work, which could be transposed in this paper as a reflection of the proposed multi-dimensional classification, in the following we relate our component model to relevant samples of state-of-the-art component models.

Gößler and Sifakis (2002) focuses on integration of components with heterogeneous interactions and execution paradigms. Our work instead aims at the integration of components implemented in different programming languages and targeting different execution platforms. The framework aims at correct-by-construction design by achieving component composability and compositionality. That work later evolved in the conception of the BIP framework (Behaviour, Interaction, Priority) (Basu et al., 2006). Separation of concerns is central to that work – though with other goals than ours – as components are created as superposition of three layers: (i) a lower behaviour layer; (ii) an intermediate layer that describes with a set of connectors the interactions between transitions of the behaviour; and (iii) an upper layer with a set of priority rules to determine the scheduling policies for interactions. The product of two components is the result of the separate composition of their layers.

The BIP framework provides atomic components, which are the basic building blocks of the system, and allows creating composite components, which are obtained by successive composition of their constituents. Notably, the authors describe the operational semantics of BIP, and an infrastructure to generate C++ code from BIP systems and an execution platform to run it either using a multi-threaded execution (each atomic component has its own thread of control) or using a single threaded execution (the execution engine is the only thread).

The two major contributions of BIP are the modeling of heterogeneous systems and the overall approach that reduces the gap between the analyzed system and the implementation. The BIP framework however does not make provisions for property preservation at run time, which is necessary to achieve *composition with guarantees,* contenting itself with semantic assurance at specification level.

SaveCCM (Hansson et al., 2004) targets heavy vehicular systems. That component model supports both time-triggered and event-triggered activation events and its components are hierarchical. Components are exclusively passive units (hence they do not comprise threads); however their description is implicitly carrying non-functional semantics: the choice of equipping a component with a trigger, data or trigger-data port implicitly dictates the concurrent semantics of the component. The ports of a component can be decorated with quality attributes to feed analysis in various non-functional dimensions and code generation. Automatic model transformations can turn the design model into a representation amenable to various forms of static analysis: timing automata with tasks or finite state process models to perform model checking of properties such as absence of deadlock or perform reachability analysis.

The PROGRESS component model (ProCom) (Carlson et al., 2010) extends SaveCCM to address high-level concerns typical of early design stages of a large-scale distributed embedded system: high-level early analysis and deployment to processing units. The component model distinguishes two granularity levels for the software specification: (i) a higher level, where the system is modeled as a set of active, concurrent subsystems which communicate through message passing; (ii) and a lower level, for the specification of the internals of the subsystems. The two levels are addressed by two separate languages ProSys for the former, and

ProSAVE for the latter (i.e., an evolution of SaveCCM). ProSys high-level components are deployed to *virtual nodes,* which are then allocated to the physical architecture. Virtual nodes are logical units for the specification of budgets with respect to CPU time and memory consumptions. The approach supports early analysis on those dimensions, to assess the goodness of the overall design before proceeding with the implementation. As a notable difference with our approach, ProCom provides explicit *connectors* related to data flow (such as "data muxer", a "data or", a "data demuxer") or control flow (such as fork and join). Clocks have to be explicitly specified in the design to provide the sources for periodic triggers.

The ROBOCOP component model (Muskens et al., 2005), targets the consumer electronic domain. The authors started their endeavour by considering the aspects that require particular attention for the target domain: (i) *upgradability,* to extend the life time of devices by uploading improved version of the software; (ii) *extensibility,* to add functionalities to the device; (iii) *low resource consumption,* particularly in footprint size, due to the limited hardware capabilities of devices. (iv) *support for third-party components,* which influences the strategy for packaging of components.

A ROBOCOP component is a collection of related models which are used to trade components between parties. Those models are human-readable models (like documentation) or machine-oriented models (a simulation model, a resource model, an interface model, a security model, etc.). In each of these models it is possible to describe the attributes of the component relatively to the dimension of interest.

The ROBOCOP component is much more similar to a package that is used to share components with various stakeholders and is different from the unit of deployment of the approach (which would be an executable component). What is termed component in component-model parlance, it is called *service* in ROBOCOP, specified in IDL. A component is developed and published in a repository. At this stage, the component is still generic as it is not bound to any target platform. Components are then tailored for execution on a specific target platform, loaded on the device, registered and instantiated for execution, and finally deployed, all via model representations.

Giotto (Henzinger et al., 2001) is a progenitor in a family of time-triggered languages and tools, which specialize for control processing where deterministic time of execution is inbred to the domain culture (Henzinger et al., 2003). Each Giotto component executes a specified number of times per period, as specified in its frequency attribute, which defaults to one. If a component has frequency greater than one, its output arcs are updated more than once per iteration, but only the final update will be visible to components with frequency one, and only on the subsequent iteration. To strengthen its fitness for strictly periodic control processing, Giotto imposes the additional constraint that the frequencies of all components in a model bear harmonic relationships with each other. This requirement clearly emanates from the wish to ease the generation of the component execution schedule and back propagates to the component model.

The time-triggered nature of the Giotto model of computation is the fundamental and radical difference to the component model presented in this paper, which instead attempts to keep the component model separate from the computational model and requires the latter to be bound to the former for a particular instantiation of the system.

The Ptolemy project (Lee, 2001) studies modeling, simulation, and design of concurrent, real-time, embedded systems realized as an assembly of concurrent components. The key principle underneath the project is the use of well-defined models of computation that govern the interaction between components. The belief behind that principle is that the choice of models of computation strongly affects the quality of a system design. And for embedded systems, useful models of computation must capture well the notions of concurrency and time. The evolution of Ptolemy supported the use of multiple models of computation constructed in a hierarchy of models as in the opinion of the Ptolemy team, no single general-purpose model of computation was likely to satisfy the needs entailed in modeling a complex embedded system.

Although not a component model itself, Ptolemy has a number of similarities to the vision presented in this paper; notably, the notion that the computational model should be understood as one essential parameter of the design problem. Ptolemy however addresses that intuition from a tool environment perspective more than from the architecture standpoint argued for in this work.

THE PROPOSED COMPONENT-BASED DEVELOPMENT APPROACH

Overall process

Fig. 3 captures what we regard as the main activities related to software development in the component-oriented approach proposed in this work.

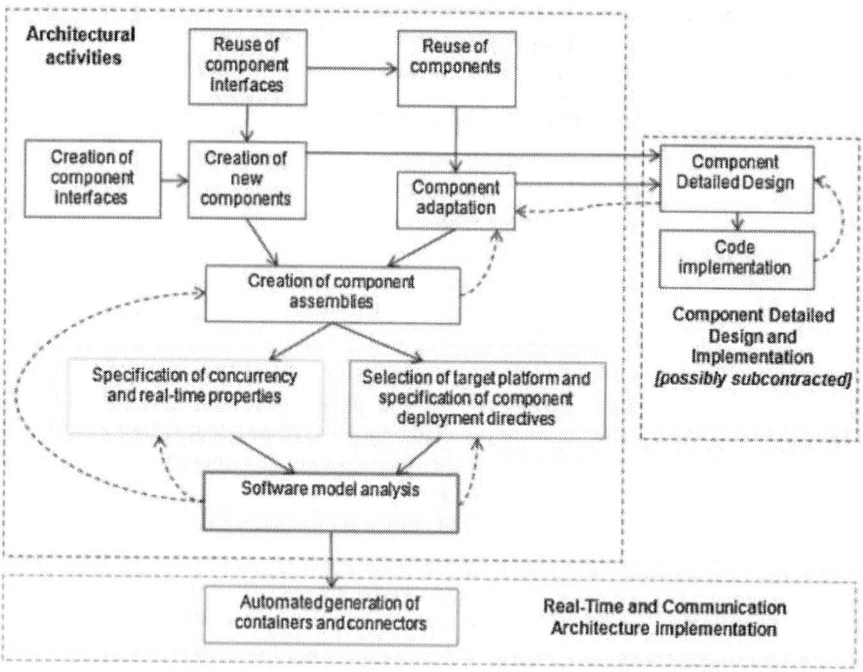

Figure 3. Our component-oriented development process.

An initial phase is concerned with the definition of components. In the proposed approach, a lot of emphasis is put on the definition of component interfaces, which exist independently of components and precede their definition (see Section 3.2).

Components can be either: (i) defined from scratch, using newly defined interfaces or reusing an interface definition; or (ii) reused from previous projects. In the latter case, if the component is not reused as is, an adaptation of the component may be required. The adaptation shall follow separation of concerns and therefore affect either interfaces alone or internals, where the latter achieves lesser reuse.

Reuse of an existing component or creation of a new component from scratch depends mainly on system requirements (functional requirements shall be compatible) and on the trade-off between the effort of producing the justification documentation for component reuse according to the applicable development process standard and the expected gain in spared development and verification and validation (V&V) effort.

Components are then bound to one another so as to create component assemblies: the complete software system is specified as a set of collaborating components.

As a distinguishing feature of our approach, all the steps described up to know are related exclusively to specification of functional concerns and functional services supported by components. All other concerns (concurrency and real-time, dependability, deployment) are addressed in separate development steps.

Fig. 3 depicts concurrency, real-time and deployment, which are directly addressed in this paper. Concurrency and real-time requirements and properties are specified by adding them on top of the functional description of services exposed by components, in the form of attribute descriptors. Deployment concerns are addressed by establishing deployment directives for each component.

The software model is then subject to analysis (e.g., schedulability analysis, bus communication analysis) to confirm that it meets the applicable non-functional requirements. Negative results from the analysis phase may require changes to the relevant non-functional attributes (e.g., periods, deadlines, priorities), deployment directives (allocation to processing units), or in the most severe cases, re-designing component assemblies or components.

Finally, containers and connectors are automatically generated using the software model as input. The former implements concurrency and real-time concerns; the latter the interaction between components. In our approach, the full real-time and communication architecture on top of the execution platform is automatically generated.

The envisioned design process recognizes two main actors: the *software architect* and the *software supplier*. The software architect represents the technical responsible for the whole software in the regard of the system-level customer. The development of part of the software may be delegated to software suppliers before its final integration.

All the activities previously described are under the responsibility of the software architect. As soon as a component is defined, it can undergo detailed design and code implementation. Those activities may of course highlight incomplete or bad definitions of components (e.g., lack of an operation in the component interface), thus requiring a renegotiation of the component definition. In any case, if an iterative/incremental development process is adopted, a number of iterations from component definition to detailed design have to be anticipated, according to the new functional perimeter of the iteration. Component detailed design and implementation are performed by software developers or may be subcontract to a software supplier.

Section 3.2 discusses those activities.

In order to organize the development process in a manner that enforces separation of concerns in the design space, we adopt the concept of "*design view*".

The ISO standard 42010 (ISO/IEC/(IEEE), 2007) stipulates that "*architectural description of the system is organized into one or more constituents called views"*, and a view is a partial representation of a system from a particular viewpoint, which is the expression of some stakeholders' concerns. If during the construction of a development approach we ratify that each view is the expression of a *single* concern, then views become effective means to enforce separation of concerns in the specification of the software system.

Design views are used mainly for two purposes: (i) view-specific visualization of entities; (ii) control of the development capabilities attributed to the designer.

As each view captures different concerns (according to the viewpoint of different software specialists), views shall enable selective visualization of only the relevant entities, and shall support creation and modification rights on those entities only. A design view can also be expression of a defined development stage. Hence it may be possible to activate it only after a certain number of conditions hold.

In the general case, views shall not incur overlaps of responsibility for creation or modification of new modelling entities. While multiple views can have read rights over cross-cutting aspects, only a single view can have create/write rights on them.

Section 3.3 provides a description on the design views defined in the proposed development process.

Design entities and design steps

The proposed development method involves the use of entities that can be differentiated between design space entities and entities of the real-time architecture. The design-level entities are explicitly specified in the design space, and require intellectual contribution from the user. The real-time architecture entities are neither specified nor explicitly represented in the design space; instead, they are automatically generated. The following entities belong in the design space: (1) data types, events and interfaces; (2) component types; (3) component implementations; (4) component instances; (5) component bindings; (6) entities for the description of the hardware topology and the target platforms. The remaining entities belong to the real-time architecture: (7) containers; and (8) connectors.

Automated generation of containers and connectors is assuredly feasible on the condition of adopting a given computational model and execution platform: that information makes it possible to define and deploy deterministic transformation rules from design-level entities to the real-time architecture formed by containers and connectors (cf. e.g., (Bordin and Vardanega, 2007), for our reference choices on this topic).

The development process proceeds across distinct design steps, some of which can be delegated to software suppliers. Easy and clear delegation of software development simplifies the technical and contractual relationship between the consortium of companies producing the software, and mitigates the relevant project risks. This is particularly relevant in ESA projects, because of *geographic return policy* considerations.[4]

We can now proceed to illustrate the sequence of design steps.

#01 – *Data types and events.* Data types are the basic entities in our approach. The designer can define a set of data types such as primitive types, enumerations, ranged or constrained types, arrays or composite types (like the *struct* in C or the *record type* in Ada). An Event is a type used for signal-based asynchronous notifications. It can comprise event parameters, which shall be typed with a datatype.

#02 – *Interfaces.* An interface is a set of one or more operations, whose signature is determined by an operation name and an ordered set of parameters, each with a direction (*in, in out, out*) and a parameter type chosen between the defined types. The interface can also defines a set of *attributes*, which are typed parameters that can be accessed through the interface. They can be read-only or read-write.

In Fig. 4 we represent three data types and an event. Interface AOCS_IF and THR_IF implement only operations while interface GYR_IF comprises one read-only attribute.

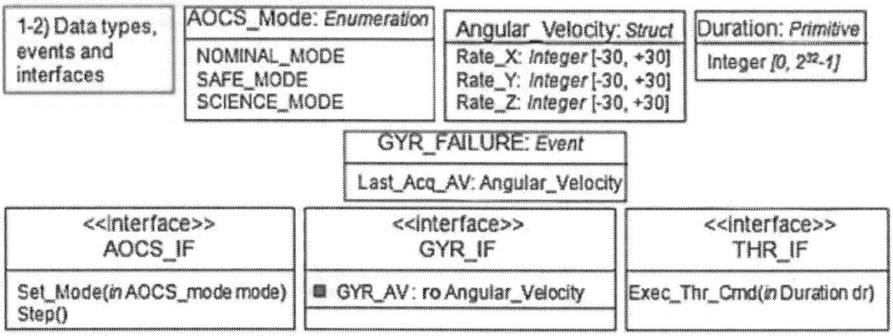

Figure 4. Data types, events and interfaces.

#03 – Component types. The component type (Fig. 5, top) is the entity that forms the basis for a reusable software asset. The software architect specifies a component type to provide a specification of the functional services of the component. The component type is specified in isolation, with no relationship with other components.

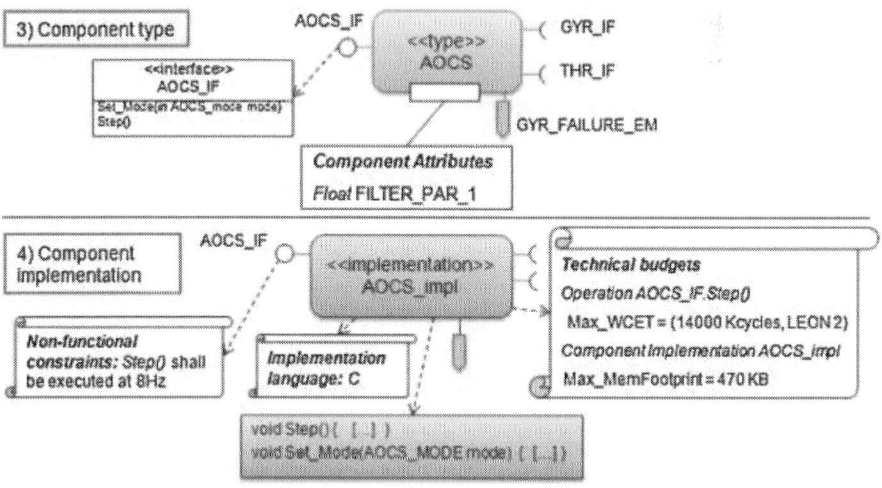

Figure 5. Component type and component implementation.

The component type therefore specifies provided interfaces (PI) and required interfaces (RI) by referencing already-defined interfaces.

The component type can also define component attributes, which similarly to interface attributes are typed parameters. Component attributes are however local to the component and cannot be accessed from the outside. They are typically used to define internal configuration parameters.

Finally, a component type may raise or receive events, via event emitter ports and event receiver ports, respectively.

In Fig. 5 (top) we defined an "Attitude and Orbit Control" (AOCS) component that provides an interface (AOCS_IF) for executing basic and application mode management functions (operation Step and Set_Mode, respectively). The component requires the interfaces GYR_IF and THR_IF, in order to perform an acquisition on a gyroscope and send pulse commands to thrusters. Finally, the component can raise events of type GYR_FAILURE, to notify that it has detected a failure of the gyroscope.

#04 – *Component implementations.* The software architect then proceeds by creating a component implementation from a component type.

A component implementation (Fig. 5, bottom) fulfils two roles: (i) it is a concrete realization of a component type; and (ii) it is the subcontracting unit of the approach, whose realization can be delegated by the system architect to a software supplier.

A component type may have several implementations (one more precise yet more computationally expensive, one more robust, etc.).

A component implementation must implement all the functional services of its type in the example above, the operations Step and Set_Mode, and includes the sequential/algorithmic code of those services and the necessary packaging information. The code is purely sequential code and shall be void of any tasking or timing constructs. Implementations can be developed in different implementation languages (i.e., Ada, C and C++ are currently supported).

Despite the sequential nature of the code, an implementation may set specific *non-functional constraints* to preserve the functional correctness of its behaviour. For example, a control law algorithm may work correctly only if executed at a certain frequency, say 8 Hz. In that case, the functional code propagates some implicit non-functional constraints which we represent in the component interface.

Additionally, the component implementation shall implement means to store the attributes exposed through its provided interfaces and those defined in its component type. The former shall be accessible via appropriate getters and setters from the interface where they are defined.

A number of technical budgets can be placed either on operations or on the whole component. The implementation of the component and its operations shall respect the allocated budget, which shall then be considered as an implementation requirement. The types of technical budget of interest emerged in our discussions with stakeholders comprise: a worst-case execution time (WCET) bound for a certain operation; a maximum memory footprint for the component implementation; maximum numbers of calls to a certain operation of a RI. The latter implicitly bounds the communication budget allowed for an implementation, as the size of data types involved in the communication are known.

This is especially useful when the software integrator wants to subcontract part of the software. After establishing how the component functionally relates to the system (by declaring its needs and provisions at the component-type level), they can derive a component implementation, attach technical budgets to it and then delegate the source code implementation to a software supplier.

It is important to notice that component implementations can (and hopefully will) undergo a detailed design activity, which may add internal decomposition (package hierarchy) or operations private to the component. This decomposition can be either be performed directly in the source code of the component implementation, or using separate implementation models (e.g., in UML). The component implementation will however be considered as a black box, as for what concerns the functional aspects, only its external interfaces matter (the same PI and RI of its component type).

For the purposes of analysis instead, we need to know which required interfaces (RI) are requested by each provided interface (PI) of a component implementation, and, more precisely, which operations in the RI are called by each operation in the various PI of the component implementation. This information is defined at implementation level, as it does not vary across different instances of the same component implementation and can be specified with a UML activity diagram or similar formalism.

#05 – Component instances and component bindings. A component instance is instantiated from a component implementation.

A component instance serves three purposes: (i) it is subject to composition with other components, and as such it is expression of functional concerns; (ii) it is the deployment unit of the approach, whereby it is expression of deployment concerns; (iii) it is the entity on which non-functional attributes are specified, hence it is expression of non-functional concerns. In this design step, we concentrate only on aspect (i).

Component bindings are set at design time between one required and one provided interface of component instances (1-to-N connections for multicast are disallowed). The binding is subject to static interface type matching to ensure that the providing end fulfills the functional needs of the requiring end.

At this stage, it is also possible to trace bindings between an event emitter ports and one or more event receiver ports. Contrarily to a binding between a RI and a PI, which establishes a functional dependency, an event binding simply establishes that the receiver component instance is interested in receiving notification of occurrences of a given event. Event-based communication is based on a publish-subscribe communication model.

In Fig. 6, a binding between one RI of the Mode_Manager component instance and the AOCS component instance is traced. This allows the Mode_Manager instance to call the operations of the interface AOCS_IF provided by the AOCS instance. The

Mode_Manager instance will also receive notification of any event of type GYR_FAILURE raised by the AOCS instance.

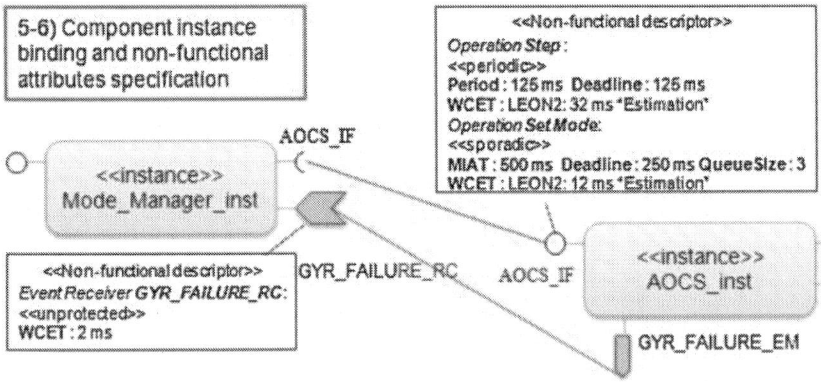

Figure 6. Component instance, component bindings and decoration with non-functional attributes.

#06 – *Specification of non-functional attributes.* After component instances have been created (Fig. 6), the software architect can add non-functional attributes to the services of their provided interfaces; those non-functional attributes are specified in non-functional descriptors.

At this stage the software architect shall specify *timing* and *synchronization* attributes. At first they shall establish the concurrent kind of the operation, by classifying it as either*immediate* or *deferred*. In the former case, the operation is executed by the flow of control on the caller side. In the latter case, the operation is executed by a dedicated flow of control in the callee. Either option has direct repercussions on the implementation of the component's container.

A stateful immediate operation is said *protected* if its execution requires to be protected from the risk of data races from concurrent calls. Otherwise it is *unprotected* and can be either stateless or otherwise free from such risks.

For a deferred operation, the architect must choose the release pattern, which can be periodic, sporadic or bursty.

- Periodic operations are executed by a dedicated flow of control released by the execution platform with a fixed period.
- Sporadic operations are executed by a dedicated flow of control which responds to a request posted via software invocation by another component or via interrupt. Two subsequent releases of the operation are to be separated by a minimum timespan called minimum inter-arrival time (MIAT). The execution platform - and not the user or the component implementor - guarantees the MIAT by enforcement. Sporadic operations require the creation of a finite-size buffer for the incoming requests, which is managed by the container (storing and fetching of requests is totally transparent to the component).
- Bursty operations, which are used to model dense releases of sporadic jobs possibly followed by spans of inactivity. There is a maximum number of activations of a bursty operation in a bounded interval; the designer is expected to provide both of them as attributes, and the size of the buffer for incoming requests, similarly to sporadic operations.

For all deferred release patterns, the software architect shall also specify the relative deadline for the completion of the operation. At a preliminary stage of development, an estimate of the worst-case execution time (WCET), based on experience from previous projects is provided, in order to enable early analysis of the overall system. The value for the WCET can later be refined with bounds for a given target platform obtained by timing analysis (Wilhelm et al., 2008).

Table 1 presents the syntax for the current set of non-functional attributes (related to concurrency and real-time) applied to the interfaces of component instances.

Table 1. Syntax in EBNF for non-functional attributes attached to operations.

EFDescriptor	= operationName, concurrencyKind;
concurrencyKind	= immediate | deferred;
deferred	= cyclic | sporadic | bursty;
immediate	= protected | unprotected;
cyclic	= "cyclic", period, WCETdesc, [deadline], [offset];
sporadic	= "sporadic", MIAT, WCETdesc, [deadline], queueSize;
bursty	= "bursty", boundedInterval, maxActivations, WCETdesc, deadline, queueSize;
protected	= "protected", WCETdesc;
unprotected	= "unprotected", WCETdesc;
WCETdesc	= execPlatform, WCETentry;
execPlatform	= "execPlatform", identifier;
WCETentry	= "WCET", naturalNumber, unit, WCETkind;
WCETkind	= "Estimation" | "Measured" | "AnalysisBound";
period	= "period", positiveNumber, unit;
deadline	= "deadline", naturalNumber, unit;
offset	= "offset", naturalNumber, unit;
MIAT	= "MIAT", naturalNumber, unit;
boundedInterval	= "boundedInterval", positiveNumber, unit;
maxActivations	= "maxActivations", positiveNumber;
queueSize	= "queueSize", positiveNumber;

It is also possible to specify end-to-end requirements on call chains across components, which is particularly useful for system-level analysis. Schedulability analysis (in the simplest forms of response-time analysis equations derived from Joseph and Pandya (1986)) can in fact provide response times only for individual tasks. The specification of end-to-end requirements can instead be used by a more expressive analysis framework (such as (Palencia and González Harbour, 1998)) to compute the completion time of end-to-end chains of operations as well as on single intermediate operations in the chain.

The component model also provides the user with means to define measurement units (and conversion factors between them). The definition of those units can then be factored out in a library reused across projects. Non-functional attributes and technical budgets that represent a dimensioned attribute (e.g., a timespan) require the specification of a value and a measurement unit.

The design environment checks that the attributes defined at instance level are compatible with the applicable non-functional constraints and technical budgets defined at implementation level.

In the example of Fig. 6 the design environment ensures that the frequency of operation*Step* matches the corresponding constraint (execution at 8Hz) in the component implementation and that the WCET is within the stipulated bound (cf. Fig. 5).

#07 – *Hardware topology and target platforms.* The hardware topology (see Fig. 7) provides a description of the system hardware limited to the aspects related to communication, analysis and code generation.

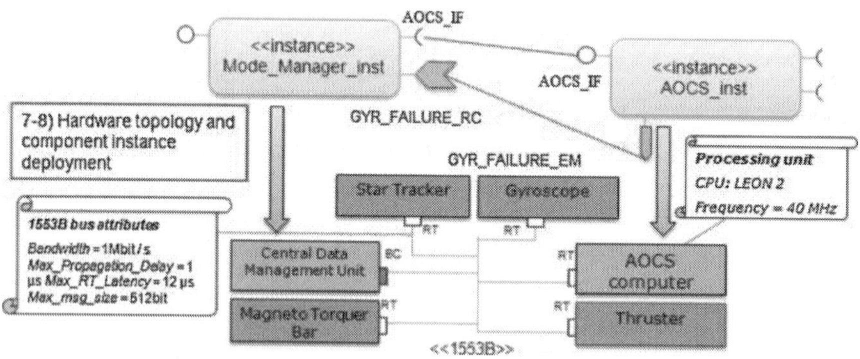

Figure 7. Deployment of component instances to processing units.

The following elements are described: (i) *processing units*, i.e., units that have general-purpose processing capability; (ii) *avionics equipment*, i.e., sensors, actuators, storage memories and remote terminals; (iii) the *interconnections* between the elements above, in terms of buses, point-to-point links, serial lines.

For the specification of those elements we use the following attributes:

- for processors: the *processor frequency,* which is used to re-scale WCET values expressed in processor cycles.
- for buses and point-to-point links: the *bandwidth,* the *maximum blocking* incurred by a message due to non-preemptability of lower-priority message transmission, the*minimum and maximum packet size,* the *minimum and maximum propagation delay,* the maximum time necessary for the bus arbiter/driver to prepare and send a message on the physical channel; and the maximum time to make it available to the receiver after reception at the destination end.

#*08 – Component instance deployment.* Once the hardware topology has been defined, the last step to perform in the design space is the allocation of component instances to processing units (see Fig. 7).
#*09 – Model-based analysis.* The system model is submitted to static analysis in the non-functional dimensions of interest. For example, schedulability analysis verifies whether the timing requirements set on interfaces can be met (Bordin et al., 2008).

The extraction of information from the user model (i.e., generation of intermediate models such as a Platform-Specific Model, PSM, or a Schedulability Analysis Model, SAM) and generation of the input for the analysis tools are automated and the results of the analysis are seamlessly propagated back to the design model as read-only attributes of the appropriate design entities.

As the model transformation that generates the model representation for containers and connectors in the SAM (see Step #10) is an integral part of the analysis transformation chain, the transformation can add information about the cost in time and space of the containers and, in particular, of connectors. This is of utmost importance for an accurate analysis of the overhead introduced by the use of platform services for local execution and for remote communication.

The analysis can be iterated at will until the designer is satisfied.

#10 – Generation of containers and connectors. Containers and connectors (Fig. 8) are generated with rules that specify: (i) the structure of each container in terms of the interface exposed to the software system, its internal threads and its protected objects; (ii) the structure of each connector; (iii) how non-functional attributes and deployment determine the creation of containers and connectors and how component instances and their operations are allocated to them.

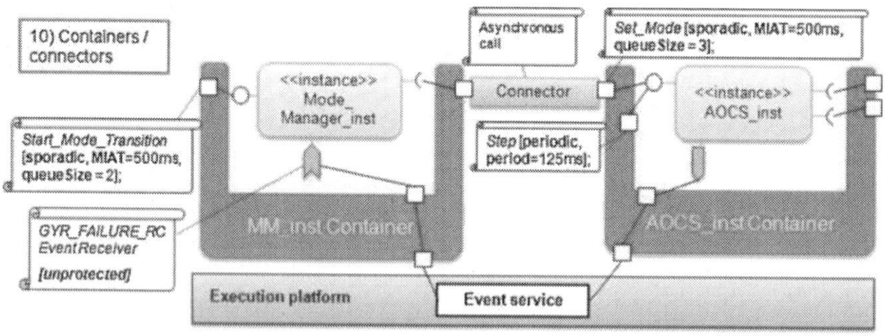

Figure 8. Automated generation of containers and connectors.

For our choice of computational model (i.e., RCM), we defined the whole set of allowable containers and connectors in a library of code archetypes (Panunzio and Vardanega, 2012), which vastly simplifies automatic code generation.

In a single-core processor setting such as ours (multi-processors systems are supported, but multi-core processors are not supported yet), implementation of concurrency is achieved with the encapsulation of sequential procedures (of component implementations) into tasks generated into containers, and the necessary protection from data races in access to shared logical resources stems from the attachment to them of concurrency control. All of that can be attained without modifying the functional logic in the relevant application code, simply by following the use relations among components and by generating the code patterns associated to the non-functional attributes of the relevant provided interfaces.

Interaction with the environment. Embedded software interacts with the environment by sampling information of interest through sensors and by controlling the plant where it executes by using actuators.

At design level we must be able to: (i) represent those devices in the hardware architecture; (ii) associate software components to them. For the latter, we use a component-level description of sensors and actuators in order to bind software components (at instance level) with the devices that they command.

The device representation is only for interaction purposes, as we only intend to represent the functional interface of the device, not its internals.

Design flow and design views

A component model does not only prescribe the syntactic rules to create design entities and how to relate them to one another, but - whether intentionally or implicitly, in any case inevitably - it also establishes a defined design flow.

The design flow comprises a series of steps that must be followed to create components, reuse components, assemble them and ultimately produce the software system. It may also determine precedence relations between those steps.

This implies that the developers of the component model shall pay careful attention to ensuring that the design flow promoted by their approach is compatible with the development process in use with the concerned industrial domain.

One of the advantages of design views is also to promote or enforce a certain design flow. The definition of the proposed component model is accompanied by the following design views (Fig. 9):

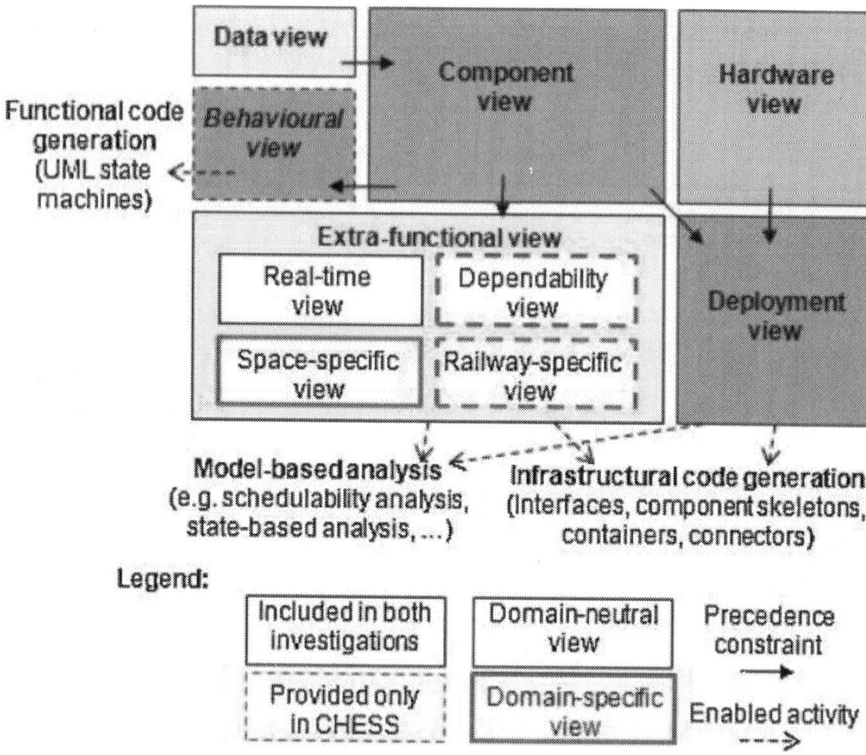

Figure 9. The design views of our component model, as defined in the two investigation strands.

- *Data view*, for the description of data types and events (which are messages generated or received by a component following a publish/subscribe model);
- *Component view*, for the definition of interfaces, components and the bindings between components to fulfill their functional needs;
- *Hardware view*, for the specification of hardware and the network topology;
- *Deployment view*, for the specification of the allocation of components to computational nodes;
- *Non-functional view*, where non-functional annotations are attached to the functional description of components. The view is divided in two sub-view according to the non-functional

concerns of interest: the real-time view and the dependability view (this is presently defined only in CHESS);

- *Space-specific view*, where the designer can specify the use of services related to*commandability* and *observability* of the spacecraft (i.e., the PUS services (European Cooperation for Space Standardization, 2003));

In the CHESS project we defined two additional views: a *Behavioural view*, and - for the lack of a better name - a *Railway-specific view*. In the former, the designer can specify in the model the functional code of a component implementation using UML state machines and generate C++ code for it (Ciccozzi et al., 2011); otherwise, the designer can always associate Ada or C/C++ source code written manually to a component implementation. In the latter, a number of railway-specific remote connection concerns are addressed.

Additionally, in CHESS, study partners focused on the development of various forms of model-based dependability analysis (i.e., error modeling, state-based analysis, FMEA, FMECA) (Montecchi et al., 2011).

In the ESA side of this work, the Behavioural view was not considered of high priority, whereas the only domain-specific concerns of interest were obviously those related to space. We implemented in that view a sizeable subset of the PUS services mentioned above, in particular: monitoring and reporting of on-board parameters, raising of nominal and erroneous events on board, commanding of on-board operations from ground. It was not possible to integrate yet dependability concerns in that strand of investigation, and they were left to follow on projects.

Interestingly, no telecom-specific view was necessary, as it turned out that the technical requirements of the domain where either already addressed by the component model or - when not addressed yet - deemed of interest also for the other two domains and therefore were promoted to the domain-neutral part of the component model (e.g.,*multiplicities* for component instances and ports, which is interesting syntactic sugar to mitigate the cluttering of the design space and the burden of specification for the user).

Fig. 10 depicts the complete design flow entailed by the proposed component model, the precedence constraints that apply to it, and the allocation of design steps and related concerns to views. The definition of data types (Step #01) shall be performed in a dedicated view were it is possible to define types independently of the underlying representation of the target platforms and selecting the encoding rules for their representation.

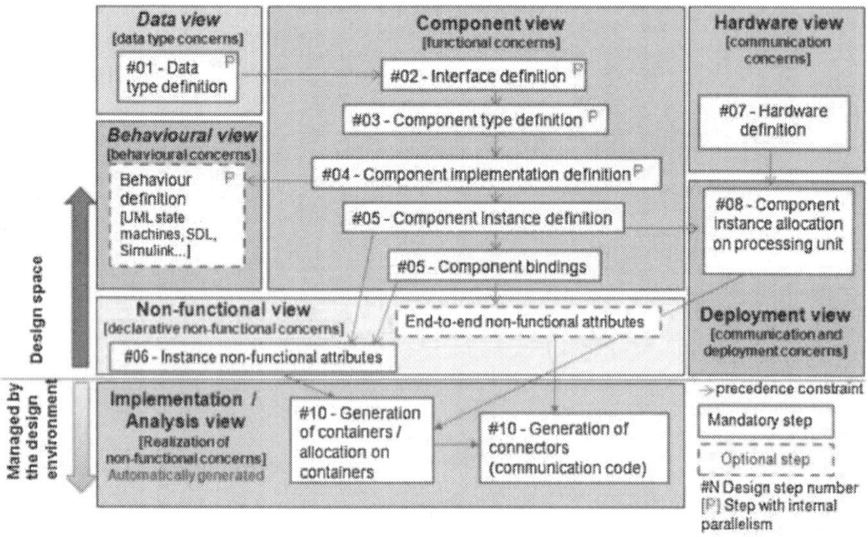

Figure 10. The design flow of our component model, including the design steps described in previous sections, and the domain-neutral design views.

The definition of interfaces (Step #02), component types (Step #03), component implementations (Step #04), component instances and instance bindings (Step #05) are allocated to a view that we termed "*Component view*" as those entities include only functional concerns.

In order to specify non-functional attributes (Step #06), we require the designer to explicitly transition to the "*Non-functional view*". In this manner, the modification rights of the latter view take effect: in the non-functional view, the designer cannot create or modify entities, but only add non-functional descriptors to the interfaces of instances. The creation or modification of entities would require the designer to return back to the "Component view".

The specification of the hardware topology and the description of the execution platform (Step #07), and the allocation of component instances to processing units (Step #08) are performed in the "*hardware view*" and "*deployment view*".

The generated containers and connectors (Step #10) realize the non-functional attributes that were specified as declarative attributes in the non-functional view. They thus belong to another view termed *implementation view*. We decided to make this view solely available in read-only mode. The reason is twofold and in line with our previous work (Bordin et al., 2008). Firstly, the view is automatically generated, so the responsibility for its correct generation is not on the designer, but on model transformation. Secondly, we do not allow manual modifications of the PSM so that we ensure consistency between the PIM and PSM levels. A similar reasoning applies also to all the necessary analysis views (for the generation of the SAM or other analysis-specific models).

Versioning, configuration management and traceability

The component model presented in this approach does not cater for versioning of interfaces or components (for example by adding annotations with version number or additional information). This choice is intentional as we consider versioning as an issue pertaining to the development life cycle and to all the artifacts produced as part it. Standard software configuration management systems (e.g., Git, SVN, Clearcase) can be used for model versioning. Staggered software releases following iterative development can be easily managed at model level with the use of branches and tags. We equally maintain that traceability of model entities (interfaces, components, etc.) to requirements shall also be managed externally (without embedding any information in the component model itself). We consider configuration management of the software model and traceability as orthogonal aspects, that shall be managed according to the applicable per-project policies. Additionally, external management of those aspects allows to address easily the reuse of components in a different project context. In fact, any versioning or traceability information embedded in the software model would be moot if a component is reused in a project under a different requirements baseline.

IMPLEMENTATION AND EVALUATION

The concepts presented in this paper, from the component model to the associated development process were the subject of two parallel prototype implementations and to a rather comprehensive range of industrial use cases.

One implementation happened within an ESA doctoral-level program, where one of the authors enrolled. That effort resulted in a domain-specific metamodel (named SCM, short for "Space Component Model") and a dedicated graphical editor based on Obeo Designer.[5] Design views were implemented using the concept of "Viewpoint" provided by the Obeo Designer framework. The use case for this effort took place in the ESA-funded COrDeT-2 study[6] and involved a thorough industrial evaluation of the component model and the associated design environment.

The other implementation took place in the ARTEMIS JU CHESS project, and retained exactly the same methodology and component model as endorsed in the parallel ESA initiative. In CHESS, the specification language for components in the PIM user model comprises UML entities, some high-level MARTE stereotypes (Object Management Group, 2009), and a few CHESS-specific stereotypes. The PSM/SAM model generated for analysis and code generation comprises exclusively MARTE stereotypes. The CHESS design environment is based on Papyrus, an Eclipse-based UML editor. A set of CHESS-specific plug-ins extend Papyrus to provide support for design views (visualization and modification rights on entities and view-specific palettes and constraints) and user-friendly automation capabilities. Model-based schedulability analysis is provided via integration with a plug-in by the University of Cantabria, which extracts the analysis information from our PSM to feed the MAST analysis tool (González Harbour et al., 2001). The results of the analysis are first propagated back to the PSM and then to the PIM, for the user to consider.

CHESS enjoyed the presence of end users from a variety of industrial domains, which included telecom, railways and space again, though from a different team than for the ESA study.

The exposure to such a score of end users allowed exposing the vision presented in this paper and its proof-of-concept technology to a rich and complementary set of case studies. Even though every individual evaluation was comparatively contained in size and effort, their total quantity caused very comprehensive coverage of all evaluation criteria.

Moreover, the very existence of the parallel implementations undertaken for ESA and for CHESS arguably shows that the proposed approach is feasible in both methodology and technology, and incurs no major difficulty in being pursued with two different specification languages and tool chains.

In the following we briefly present the essence of each case study, while in Section 4.5we summarize the results and the feedback we received from them.

ESA: reference Earth Observation case study

The first case study was performed in the scope of the COrDeT-2 study. It concerned the re-engineering of a small yet representative Earth Observation mission, used as a reference in many R&D studies at ESA. It consists in a small satellite in Low Earth Orbit, comprising an optical payload to capture images. The mission was originally developed with a traditional code-centric approach.

In this case study, a subset of the on-board software was re-designed, basing on the software reference architecture approach and using the supported component model.

This case study focused on: (1) ensuring that the component model is able to express the needs for the development of on-board software for satellites; (2) ensuring that the component model is able to accommodate space-specific needs (namely, the "Packet Utilization Standard" (PUS) services for commandability and observability of the spacecraft from ground stations) in a manner that is consistent with the rest of the approach, and avail to the designer specification means at the right level of abstraction.

The case study was performed by a senior engineer with no prior experience on component-oriented methodologies, from a small-size on-board software prime contractor, with support by a part-time consultant who participated in the ESA investigation. Occasional support by one of the authors on some details of the methodology was also needed. The case study spanned 6 months. The conclusions of the investigation were reviewed by two large software and system prime contractors of the domain.

CHESS: space case study

The CHESS space use case was based on Sentinel-3, an Earth observation mission within ESA's Living Planet Program. The use case modeled a sizeable subset of the on-board software of the Sentinel-3 satellite: AOCS (Attitude and Orbit Control System), EM (Equipment Management), PM (Platform Management, an abstraction layer between the SW applications and platform resources such as the 1553B command bus, and the on-board time reference), TR (Thermal Regulation), SADM (Solar Array Drive Mechanism).

This case study focused on: (1) ensuring that the CHESS methodology is compatible with the current process and practices of the domain stakeholder; (2) code generation of containers and connectors, by means of a model transformation towards a proprietary modeling infrastructure (MyCCM); (3) verifying the non-functional properties of the model - in particular timing properties - utilizing the supported model-based analysis and the back-propagation mechanism to rapidly iterate the analysis.

The case study was performed by a R&D engineer with experience on component-oriented methodologies in two iterations: a shorter one of approximately 1 month and a second one that spanned 2 months.

CHESS: telecom case study

The telecom use case was based on the "Connectivity Packet Platform" (CPP), which allows constructing packet access nodes based on IP and ATM transport technologies. It provides cluster functionalities, redundancy, fault tolerance, and can be considered as a soft real-time system with a few components with stringent time requirements.

This case study focused on: (1) assessing the use of the component model for their development process; (2) modeling of functional code via ALF state machines; (3) functional code generation from state machines. The generated code is in C++ and targets their reference execution platform;

The case study was performed by two junior engineers with support by a senior engineer for the integration of their reference execution platform in the approach. It was performed along a time-span of approximately 9 months, in two iterations. Some work was re-done in the second iteration, after the team had gained better understanding of the methodology.

CHESS: railway case study

The railway case study was based on applications related to the European Rail Traffic Management Systems (ERTMS).[7] ERTMS comprises two main constituents: (i) the ETCS (European Train Control System), which is used to transmit information to the train driver (train speed, calculation of breaking curves,...) and to monitor the compliance of the driver with prescriptions; (ii) the GSM-R standard, to enable bi-directional wireless communication exchanges between the ground system and the train. The case study concerned a commercial solution for the monitoring and analysis of the strength of the up-link and down-link GSM-R signal (respecting bandwidth constraints) in proximity of a high-speed/high-capacity railway line. It analyses possible interferences on the signal and can discriminate if the interference originates from outside the train or on board; in the latter case the train driver is notified, so as to take appropriate actions.

The case study focused on modeling two subsystems: (i) an "analyzer", which performs the analysis of the GSM-R signal; (ii) a "receiver", which receives the analyzed data on signal quality on board and performs the appropriate actions in response.

The "analyzer" is deployed on a laptop, the "receiver" on a dedicated board (simulated with a laptop in the case study).

The case study was small-sized, yet centered on several key features: (i) support for the creation of components written either in C or Ada; (ii) support for multi-node systems; (iii) support for a railway-specific communication protocol to regulate communication from (i) to (ii).

The case study was performed by an engineer without previous experience in component oriented methods, with support on dependability modeling from a lead engineer. The case study spanned along two iterations over a total period of 4 months. Considerable technical support by one of the authors was necessary during the second iteration.

Summary evaluation

The overall result of the above evaluations is reported in Table 2. In the following we briefly summarise the feedback obtained from the industrial users in the three main dimensions of their evaluation: the fitness of the development process for the industrial domain; the quality of the product as resulting from the chain of model transformations; the viability of use in production.

Table 2. I (Demonstrated and considered interesting and promising); A (Demonstrated and considered adequate); H (High); M (moderate); L (low).

ID	Result/criteria short name	ESA case study	CHESS: telecom	CHESS: railway	CHESS : space
PR-1	Adoption of design views	I	I		I
PR-2	Component-oriented design process	A	A	A	A
PR-3	Automated generation of non-functional code			I	I
PD-1	Containers for separate realization of non-functional properties	I		I	I
PD-2	Support for multiple target platforms	A	A	A	A
M-1	Maturity of methodology	H	H	H	H
M-2	Increase of productivity	M	M	L	L
M-3	Learning curve	H	H	M	H

Process-related aspects

The adoption of design views (PR-1) as a means to enforce separation of concerns is cited as an important factor in the telecom and space domain feedback of the CHESS project and in the ESA investigation.

The component-oriented development process that was defined (PR-2) is cited by all case studies as adequate for the realization of their target system. It helps to split the intellectual work in manageable parts that can be realized (or further refined) independently by the designers. Furthermore, it is considered as an enabler for efficient multi-team development.

The automated generation of non-functional code, plus the interface code and the skeletons for components (PR-3) was successfully achieved in the railway case study of CHESS. The code generation was also evaluated by the CHESS space case study, and deemed satisfactorily and promising.

Product-related aspects

Generation of containers and connectors (non-functional code) from functional code (PD-1) was demonstrated in the CHESS space and railway use cases and in the ESA investigation, and considered an important achievement.

Support for multiple target platform (PD-3) was demonstrated by the support for the target languages (Ada, C, C++) and reference execution platforms selected by the represented application domains. It confirms the goodness of fit of the approach for a cross-domain exploitation.

Miscellaneous aspects

The maturity of the methodology (M-1) is considered high by the feedback of all case studies.

The gain in productivity of the approach (M-2) is still considered low to moderate. This feedback is mostly due to the prototypical nature of the two toolsets and to the lack of collaborative features for the modeling activity. We however are satisfied by this fair evaluation, as the feedback by industrial users highlights how this is exclusively a technological problem (and in fact the methodology itself was considered mature).

Finally, the learning curve of the approach (M-3) is considered from moderate to high. This highly depends on the previous exposure of the project partner to MDE and component-oriented approaches (the CHESS space partner is quite familiar with those paradigms, the space partner of the ESA investigation and the telecom and railway partners of CHESS were not). This highlights in particular: the difficulty on fitting a novel development approach into the pre-existent industrial process of a stakeholder, which is an aspect that is seldom ignored or underestimated, especially in academic literature. It also highlighted that an increasing effort shall be devoted to dissemination activities, especially in the form of tutorials or reference guides (for example, with the description of architectural patterns for solving recurrent design problems using the component model and the reference architecture).

CONCLUSIONS AND FUTURE WORK

In this paper we presented a novel component model developed in the context of an investigation of the European Space Agency that aims at the definition of a software reference architecture for the on-board software of future ESA mission and in the parallel multi-domain project CHESS.

The component model was designed to support separation of concerns, in particular between functional and non-functional concerns. Separation of concerns is achieved with careful allocation of distinct concerns to the software entities of the approach: components, containers and connectors. Our components are void of any non-functional concern such as - in the dimension of concurrency - tasking, synchronization, timing. This ensures that they can be directly reused in different contexts under different non-functional requirements.

In our approach, component types and implementations only address functional concerns. Non-functional attributes are later superimposed on component instances. No premature non-functional choice is in fact forced on the component design. Furthermore, all the non-functional decisions are postponed to the appropriate design stage and no component description or implementation requires any modification in case non-functional attributes are changed.

After non-functional attributes have been set, a call chain may span across components but it is guaranteed to be *finite* (it is represented as a tree with a leaf at every first occurrence of a deferred operation, i.e., with sporadic or bursty activation pattern, which is executed by its own thread of execution) and *acyclic* (Cancila et al., 2010).

Finally, it is worth of notice that in our approach, software suppliers *solely* address functional concerns (i.e., the implementation of the sequential/algorithmic code), under a defined budget envelope negotiated with the software architect. The specification of any non-functional concern (*how* the software shall be executed) stays the responsibility of the software architect; and the implementation of non-functional concerns stays the responsibility of the design environment (through automated code generation).

We also support model-based analysis of non-functional concerns to help the designer evaluate the design in early phases of development. The results of the analysis are directly presented as attributes of the entities of the user model.

The two implementations of the proposed component model that we developed attest the methodological goodness of the approach withstanding the use of different specification languages and technologies.

We are currently extending the component model described herein to support hierarchical components. Hierarchical decomposition will be used mainly to master the design complexity and contain the cluttering of entities in the design space: a problem common to all MDE approaches. Hierarchical composition will be used to aggregate reused components (fetched from a component repository). We are also carefully revisiting the design stages when hierarchical components are used, so that they fit our design process and do not break the separation of concerns principle.

ACKNOWLEDGEMENTS

This work was supported by the Networking/Partnering Initiative of the European Space Agency and by the CHESS project under ARTEMIS JU grant no. 216682.

REFERENCES

1. Basu, A., Bozga, M., Sifakis, J., 2006. Modeling heterogeneous real-time components in BIP. In: Proceedings of the 4th IEEE International Conference on Software Engineering and Formal Methods, pp. 3-12.
2. Bordin, M., Vardanega, T., 2007. Correctness by construction for high-integrity real-time systems: a metamodel-driven approach. In: Proceedings of the 12th International Conference on Reliable Software Technologies - Ada-Europe.
3. Bordin, M., Panunzio, M., Vardanega, T., 2008. Fitting schedulability analysis theory into model-driven engineering. In: Proceedings of the 20th Euromicro Conference on Real-Time Systems.
4. Burns, A., Dobbing, B., Vardanega, T., 2014. Guide for the Use of the Ada Ravenscar Profile in High Integrity Systems, Technical Report YCS-2003-348. University of York.
5. Cancila, D., Passerone, R., Vardanega, T., Panunzio, M., 2010. Toward correctness in the specification and handling of nonfunctional attributes of high-integrity real-time embedded systems. IEEE Trans. Ind. Inform. 6 (2), 181–194.
6. Carlson, J., Feljan, J., Mäki-Turja, J., Sjödin, M., 2010. Deployment modelling and synthesis in a component model for distributed embedded systems. In: Proceedings ofthe 36th Euromicro Conference on Software Engineering and Advanced Applications.
7. Chapman, R., 2006. Correctness by construction: a manifesto for high integrity software.
8. In: ACM International Conference Proceeding Series, vol. 162.
9. Chaudron, M., Crnkovic, I., 2008. Component-based software engineering. In: van Vliet, H. (Ed.), Software Engineering: Principles and Practice. Wiley (Chapter 18).
10. Ciccozzi, F., Cicchetti, A., Krekola, M., Sjödin, M., 2011. Generation of correct-byconstruction code from design models for embedded systems. In: Proceedings of the 6th IEEE Int. Symposium on Industrial Embedded Systems, pp. 63–66.
11. Crnkovic, I., Sentilles, S., Vulgarakis, A., Chaudron, M.R.V., 2011. A classification framework for software component models. IEEE Trans. Softw. Eng. 37 (5), 593–615.
12. Dijkstra, E.W., 1972. The humble programmer. Communications of the ACM 15 (10), 859–866, ISSN 0001-0782.
13. Dijkstra, E.W., 1982. On the role of scientific thought. In: Dijkstra, E.W.(Ed.), Selected Writings on Computing: A Personal Perspective. Springer-Verlag New York Inc., pp. 60–66, ISBN 0-387-90652-5.
14. European Cooperation for Space Standardization (ECSS), 2003. Space Engineering - Ground Systems and Operations - Telemetry and Telecommand Packet Utilization, ECSS-E-70-41A.

15. Gößler, G., Sifakis, J., 2002. Composition for component-based modeling. In: First International Symposium on Formal Methods for Components and Objects, pp. 443–466.
16. González Harbour, M., Gutierrez, J., Palencia, J., Drake, J., 2001. MAST: modeling and analysis suite for real-time applications. In: Proceedings of the 13th Euromicro Conference on Real-Time Systems.
17. Hansson, H., Akerholm, ˚ M., Crnkovic, I., Törngren, M., 2004. SaveCCM - a component model for safety-critical real-time systems. In: Proceedings of the 30th Euromicro Conference, pp. 627–635.
18. Henzinger, T., Horowitz, B., Kirsch, C.M.,2001. Giotto: a time-triggered language for embedded programming. In: Proceedings of EMSOFT 2001, vol. LNCS(2211). Springer-Verlag.
19. Henzinger, T.,Kirsch,C., Sanvido,M., Pree,M., 2003. Fromcontrolmodels to real-time code using Giotto. IEEE Control Syst. Mag. 23 (1).
20. ISO/IEC/(IEEE), 2007. Systems and Software Engineering - Recommended Practice for Architectural Description of Software-Intensive Systems, ISO/IEC 42010 (IEEE Std) 1471-2000.
21. Joseph, M., Pandya, P.K., 1986. Finding response times in a real-time system. Comput. J. 29 (5), 390–395.
22. Lau, K.-K., Wang, Z., 2007. Software component models. IEEE Trans. Softw. Eng. 33 (10), 709–724.
23. Lee, E., 2001. Overview of the Ptolemy Project. http://ptolemy.eecs.berkeley.edu/ Mehta, N.R., Medvidovic, N., Phadke, S., 2000. Towards a taxonomy of software connectors. In: Proceedings of the 22nd International Conference on Software Engineering, pp. 178–187.
24. Montecchi, L., Lollini, P., Bondavalli, A., 2011. Dependability concerns in modeldriven engineering. In: Proceedings of the 14th IEEE Int. Symposium on Object/Component/Service-Oriented Real-Time Distributed Computing Workshops, pp. 254–263.
25. Muskens, J., Chaudron, M., Lukkien, J.J., 2005. A componentframework for consumer electronicsmiddleware.In: Atkinson,C., et al.(Eds.),Component-BasedSoftware Development for Embedded Systems, vol. 3778 of Lecture Notes in Computer Science. Springer Berlin/Heidelberg, pp. 164–184.
26. Object Management Group, 2009. UML Profile for Modeling and Analysis of Real-time and Embedded Systems (MARTE), version 1.0, urlhttp://www.omg.org/spec/MARTE/1.0/http://www.omg.org/spec/MARTE/1.0/.
27. Palencia, J.C., González Harbour, M., 1998. Schedulability analysis for tasks with static and dynamic offsets. In: Proceedings of the 19th IEEE Real-Time Systems Symposium.

28. Panunzio, M., Vardanega, T., 2007. A metamodel-driven process featuring advanced model-based timing analysis. In: Proceedings of the 12th International Conference on Reliable Software Technologies - Ada-Europe.
29. Panunzio, M., Vardanega, T., 2010. A component model for on-board software applications. In: Proceedings of the 36th Euromicro Conference on Software Engineering and Advanced Applications, pp. 57–64.
30. Panunzio, M., Vardanega, T., 2012. Ada Ravenscar code archetypes for componentoriented development. In: Proceedings of the 17th International Conference on Reliable Software Technologies - Ada-Europe.
31. Panunzio, M., Vardanega, T., 2013. On software reference architectures and their application to the space domain. In: 13th International Conference on Software Reuse, pp. 144–159.
32. Puschner, P., Kirner, R., Pettit, R., 2009. Towards composable timing for real-time software. In: Proceedings of the 1st International Workshop on Software Technologies for Future Dependable Distributed Systems, as Part of ISORC.
33. Schmidt, D.C., 2006. Model-driven engineering. IEEE Comput. 39 (2), 25–31.
34. Sifakis, J., 2005. A framework for component-based construction extended abstract. In: Proceedings of the 3rd IEEE International Conference on Software Engineering and Formal Methods, pp. 293–300.
35. Spitznagel, B., Garlan, D., 2001. A compositional approach for constructing connectors. In: IEEE/IFIP Conference on Software Architecture, pp. 148–157.
36. Szyperski, C., 2002. Component Software: Beyond Object-Oriented Programming, 2nd ed. Addison-Wesley Professional, Boston.
37. Vardanega, T., 2009. Property preservation and composition with guarantees: from ASSERT to CHESS. In: Proceedings of the 12th IEEE International Symposium on Object/Component/Service-Oriented Real-Time Distributed Computing, pp. 125–132.
38. Wilhelm, R., Engblom, J., Ermedahl, A., Holsti, N., Thesing, S., Whalley, D., Bernat, G., Ferdinand, C., Heckmann, R., Mitra, T., Mueller, F., Puaut, I., Puschner, P., Staschulat, G., Stenstr”oem, P., 2008. The worst-case execution time problem: overview of methods and survey of tools. IEEE Trans. Embed. Comput. Syst. 7 (3), 1–53.

CITATION

Marco Panunzio, Tullio Vardanega, A component-based process with separation of concerns for the development of embedded real-time software systems, Journal of Systems and Software, Volume 96, October 2014, Pages 105-121, ISSN 0164-1212, http://dx.doi.org/10.1016/j.jss.2014.05.076.

CHAPTER 4

Characteristics of Software Ecosystems for Federated Embedded Systems: A Case Study

Jakob Axelsson[1,2], Efi Papatheocharous[1], Jesper Andersson[3]

[1] Swedish Institute of Computer Science (SICS), SE-164 29 Kista, Sweden
[2] School of Innovation, Design and Engineering, Mälardalen University, SE-721 23 Västerås, Sweden
[3] Department of Computer Science, Linneaus University, SE-351 95 Växjö, Sweden

ABSTRACT

Context

Traditionally, Embedded Systems (ES) are tightly linked to physical products, and closed both for communication to the surrounding world and to additions or modifications by third parties. New technical solutions are however emerging that allow addition of plug-in software, as well as external communication for both software installation and data exchange. These mechanisms in combination will allow for the construction of Federated Embedded Systems (FES). Expected benefits include the possibility of third-party actors developing add-on functionality; a shorter time to market for new functions; and the ability to upgrade existing products in the field. This will however require not only new technical solutions, but also a transformation of the software ecosystems for ES.

Objective

This paper aims at providing an initial characterization of the mechanisms that need to be present to make a FES ecosystem successful. This includes identification of the actors, the possible business models, the effects on product development processes, methods and tools, as well as on the product architecture.

Method

The research was carried out as an explorative case study based on interviews with 15 senior staff members at 9 companies related to ES that represent different roles in a future ecosystem for FES. The interview data was analyzed and the findings were mapped according to the Business Model Canvas (BMC).

Results

The findings from the study describe the main characteristics of a FES ecosystem, and identify the challenges for future research and practice.

Conclusions

The case study indicates that new actors exist in the FES ecosystem compared to a traditional supply chain, and that their roles and relations are redefined. The business models include new revenue streams and services, but also create the need for trade-offs between, e.g., openness and dependability in the architecture, as well as new ways of working.

INTRODUCTION

In many industries developing technical products, such as automotive, aerospace, or process automation, Embedded Systems (ES) and software play an increasingly important role [1].

Traditionally, the ES has been an integral part of the physical product and being in charge of controlling and monitoring the product during its operation. Some of the key characteristics of many ES are that they have to be cost-efficient, reliable, robust, safe, and secure, and hence they are typically tailored for a certain product.

The software of the ES is usually monolithic in the sense that it is handled in the product as one piece, without the possibility to replace only parts of it, or add new parts on top of the present ones. The software is installed during production, and upgrades normally require physical connection to the ES. Often, the ES is developed by an external supplier based on a specification from the manufacturer responsible for integrating the final product.

With the arrival of affordable communication technologies, in particular wireless, it is becoming feasible to provide external communication capabilities to the ES. This eventually allows updates of software to be carried out remotely even though the flexibility of such upgrades would be limited due to the monolithic nature of the ES software. Recent research [2] has however demonstrated how a software plug-in mechanism can be utilized, that allows the installation of add-on functionality in the ES and changes its monolithic nature. Such a plug-in mechanism, in combination with external communication, opens up many new possibilities for ES in particular and product development in general:

- It shortens dramatically the time to market for introducing new or extended features, and allows addition of new features into products that are already deployed, something which was not feasible before.
- It allows the creation of Systems-of-Systems (SoS), by letting plug-ins in different products connect to each other and exchange data through communication channels, thus forming a Federated Embedded System (FES), in ways that were not thought of at the time of design of the individual systems.
- It provides opportunities for third-party developers to supply add-ons, and thereby fosters open innovation formed by

openness in collaborations in a way that was never seen before in the area of ES.

As an illustrative example of a FES, let us consider the case of car-to-car and car-to-infrastructure communication. In such a scenario, one could imagine a service where information is passed to cars about the recommended speed to be able to pass road intersections without having to stop at red lights. The cars and the servers in the road infrastructure form a federation to achieve this service. The recommended speed could be fed into the car's cruise control system to automatically adapt the car's speed. Thus, the service could decrease fuel consumption and travel time, and improve traffic flow. The information gathered by this service could further be used in real-time services that estimate the travel time, but also in aggregated services that show traffic patterns for route guidance, or even for long-term road network planning improvements. It is in this combination of direct improvements to the individual ES products and the creation of novel services that the real potential of FES resides.

These possibilities lead to many challenges on a technical level. In addition, they also have large implications on how the development process is instantiated and executed, and changes the relations between the different parties involved. The supply chain model traditionally used for ES, with a manufacturer integrating parts ordered from suppliers, will need to expand into a much more dynamic software ecosystem [3], where many new actors enter the ecosystem and interact. This perspective involves expansion beyond and across organizational boundaries, exposing platforms and opening access to reusable assets, increasing collaboration in the software ecosystem, but also involving the need of management of an economic, business and social environment [4].

The contribution of this paper is an empirical study of how such an ecosystem for FES should be set up and supported, resulting in a characterization of the business aspects, development issues, and architectural strategies. This ecosystem includes the ES with its hardware and built-in software; the software plug-ins that are added dynamically to the ES; the software running on servers that is needed to create services in the FES; and also to some extent, the mechanical products on which the ES are installed. Ecosystems in the ES domain are different than typical software ecosystems because the processes carried out need to support building multiple complex units of distributed functionality with real-time properties and constraints.

Research questions

The possibility of creating FES is currently emerging, and many companies in different industries are struggling with how to position themselves in this new ecosystem. Often, it will require the companies to make major revisions to their core business strategies, as well as their technology platforms, to remain competitive. The *goal* of this research is to provide support for such decisions by providing a more complete picture of how an ecosystem for FES should be organized. The underlying hypothesis of the research is:
H.

The business related aspects are just as important as technology to the industry actors in order to make FES successful.

Therefore, two primary *Research Questions (RQs)* have been studied:

1. *What actors are needed in an ecosystem of Federated Embedded Systems (FES), and what are their relations?*
2. *What are the key elements of (a) the business models, (b) the product architecture, and (c) the processes, methods, and tools, needed to make the ecosystem effective?*

As a side effect of studying these RQs, we expected that various other relevant areas related to software ecosystems for FES would be discovered. Even though the development of FES technology is at a relatively early stage, we find it essential to have an early look at the non-technical aspects around it, since business models and development processes must be in place to make the technology effective. The way actors in the ecosystem interact will most likely also influence technology, by requiring different mechanisms to support development and operation of the system, and this research aims at discovering those mechanisms up front in order to ensure that they are properly reflected in future technical solutions.

Research method

Ecosystems for FES have only recently started to emerge, and therefore it is hard to find good examples of mature constellations to study empirically. Therefore, our research is highly explorative in its nature, and the best we can do to gain empirical data is to study similar ecosystems in related areas and try to extract and extrapolate information from there. The goal is thus to formulate a conceptual model which can be thought of as a set of hypotheses describing the important elements and relations in an ecosystem for FES. Naturally, these hypotheses have to be validated in future research.

There are several candidate research methods that could suit this kind of research, including case studies, grounded theory, and ethnographic studies. Grounded theory is useful when there is a need to generate new theories, but in this case there are already established theories for software systems, and it did not appear likely that the extension to FES would require a whole new set of concepts. As for ethnographic methods, it was difficult due to practical reasons. It would have required us to follow a certain ecosystem within it for some time, and we did not have access to a suitable object of study to allow this.

The research method we have chosen to apply is therefore an explorative case study, details of which can be found in [5]. In this research, the context is the relevant industry developing technical products and services in general. The focus is on the ecosystem as a whole, rather than on an individual company. The study includes a number of different companies, and is thus, according to the terminology of [5], an embedded case study with multiple units of analysis.

Since the study is not based on a pre-defined theory and it is performed for exploratory purposes, it is important to be open-minded and open-ended in the data collection process, and therefore a research method based on semi-structured interviews was considered suitable. In this way, an a priori idea of the most important aspects can be complemented with the spontaneous input from interviewees. This also makes the study qualitative in nature, and it has not been a target to collect substantial quantitative data.

In addition, the research method of the study serves the purpose of descriptive research, portraying the situation or phenomenon, based on Robson's classification [6]. It involves information from a specific population or some sample from one. The method is closely related to a case study and is primarily used for exploratory purposes, but also for descriptive purposes in the cases where the generality of the situation or phenomenon is or can be of secondary importance. The basic process steps in conducting case study research proposed in [7] were applied, particularly, (1) case study design, (2) preparation for data collection, (3) collecting evidence, (4) analysis of collected data, and, (5) reporting. The adaptation of these generic steps to the concrete setting was defined in a case study protocol during the design of the study. Fig. 1 shows an overview of these steps in the study, with indications on where in this paper the different parts are covered.

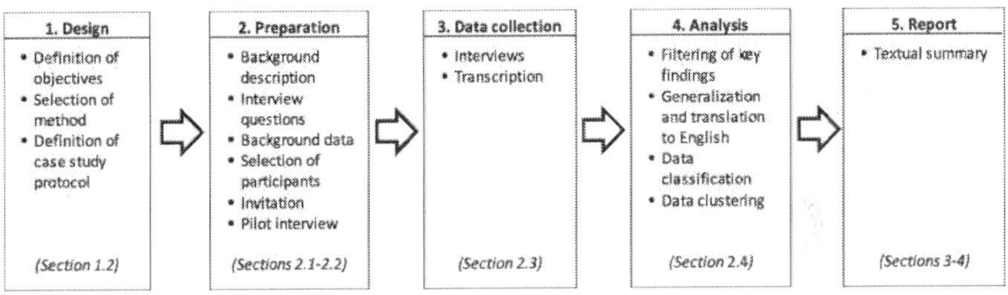

Figure 1. Overview of the study.

Overview of paper

The remainder of this paper is structured as follows. In the next section, the design of the case study is presented in more detail, including information on the interviews conducted and the analysis framework used. In Section 3, the analysis of the collected data is presented, and Section 4 discusses these findings with a focus on the areas described in the RQs. In Section 5, the validity of the findings is analyzed, and Section 6 provides a review of related research. Finally, in Section 7, the conclusions are summarized together with indications of future research.

CASE STUDY

In this section, the design of the case study is presented. The description includes what kind of preparation was done, how participants were selected and contacted, how the interviews took place, and the design of the analysis framework used to extract the relevant information from the interview data in order to address the RQs.

Preparation

In order to ensure that all relevant aspects were covered during the interviews, a set of interview questions were prepared in advance. The questions (Appendix A) served as a checklist, and also drove the interview forward in case this did not happen naturally. There were approximately 20 interview questions prepared beforehand divided in five categories: (1) Preliminary data about the interviewee and the organization; (2) roles and relations in the ecosystem (RQ1); (3) business models (RQ2a); (4) development processes, methods, and tools including product architecture (RQ2b-c); and (5) wrap-up questions, including an open question if anything relevant should be added to the discussion.

In addition, a short text of approximately half a page was prepared which was used to frame the background and objectives of the interview. The text also explained how the data would be used and included ensuring the confidentiality and anonymity of the participants and their organizations. The first interview conducted was used as a pilot, with the intention of updating the interview questions and background text for the remaining interviews. However, the pilot did not reveal any need to change the interview questions, and all interviews were conducted based on the original material prepared. Prior to the interviews, some background data was also gathered from public sources about the companies, their business and the interviewees, in the cases where they were not known.

Participants

The selection of the participants in the interviews can be best described as a stratified convenience sampling. The stratification was in the sense that we strived to include companies that we could beforehand understand would be most likely to play different roles in a future ecosystem. Therefore, we included participants from different domains, namely manufacturers, traditional suppliers, service operators, and add-on software developers. Also, we included participants that play different roles in their organization, including both engineers and business people, and both managers and non-managers. In addition, we included participants from several industrial domains, namely automotive, off-road machinery, and automation.

The sampling was convenience-based, since most participants were, in one way or another, acquainted with the researchers before. This included old colleagues, previous research partners, or just more remote and recent contacts of the researchers. We considered this necessary for two reasons: it allowed us to find suitable people in the organizations to interview, thus ensuring high quality data collected from the interviews; and it ensured that the participants trusted the researchers enough to be able to speak openly without the fear of confidential information being disclosed.

The participants were initially contacted via e-mail, receiving a short personal message, which differed depending on the participant's relation with the researchers. Attached to the e-mail was also the short text describing the background and objectives of the research, mentioned in the previous section. After a positive response had been received, an interview time was decided. In several cases, the participant asked if they could bring a colleague to the interview, and this was accepted.

We performed in total 12 interviews with 15 persons from 9 companies. Table 1 shows an overview of the interviews, companies, and participants. All of the participants but one had a university education, and the average time from graduation was over 20 years, which indicates that the selection of participants involved highly knowledgeable people. The subjects were also considered to be experienced within their companies, with an average of over 10 years with their current employer. All participants were Swedish, and act in leadership roles (i.e., line managers, project managers, technical experts and system architects). The order of the interviews in Table 1 follows the order in which they were carried out. Regarding the size of companies, it was a bit tricky to define exactly, since many companies form part of a larger group, and the size can either represent the unit or the group. In the table, the companies are classified in sizes according to the unit at the site in question, and using the scales of Small (S) (<10 employees), Medium (M) (10–1000), and Large (L) (>1000).

Table 1. Overview of interviews, companies, and participants.

Interview			Company				Participant			
Id	Duration (hh:mm)	No. of statements	Id	Industry	Role	Size (S, M, L)	Id	Years employed	Years since MSc	Gender (M, F)
1	1:10	39	A	Automotive	Service operator	L	i	3	12	F
2	1:10	27	B	Automotive	Supplier	M	ii	5	40	M
							iii	1	11	M
3	0:55	26	A	Automotive	Service operator	L	iv	10	18	M
							v	20	25	M
4	0:50	20	C	Automotive	Manufacturer	L	vi	20	26	F
5	2:00	62	C	Automotive	Manufacturer	L	vii	26	N/A	M
6	1:10	32	C	Automotive	Manufacturer	L	viii	14	20	M
7	0:50	19	D	Automotive	Manufacturer	L	ix	27	35	M
							x	10	8	M
8	1:20	41	E	Software Consultancy	Add-on developer	S	xi	1	19	M
9	0:45	21	F	Embedded Systems	Supplier	M	xii	5	16	M
10	0:57	24	G	Embedded Systems	Supplier	L	xiii	21	22	M
11	1:10	44	H	Off-road Machinery	Manufacturer	M	xiv	5	18	M
12	1:15	15	I	Automation	Manufacturer	M	xv	6	16	M
Avg	1:07	30.8						11.6	20.4	

Data collection

The interviews started by the researcher shortly recapitulating the background of the interview, based on the short text that was included in the e-mail sent to the participant. In most cases, this was sufficient for the respondents to start discussing, and there were only limited need for the researchers to pose direct questions. However, towards the end of each interview, the set of prepared interview questions (see Appendix A) was used as a checklist by the researcher to ensure that no major area intended to be discussed was left uncovered. Often, the respondents illustrated their points by giving concrete examples from their companies. All interviews were conducted in Swedish. Most interviews were carried out face-to-face, except interview 9, which was over telephone. The first author conducted ten of the interviews, and the last author the remaining two (interviews 10 and 12). One researcher thus participated in each interview and usually took handwritten notes. This method of data capturing was chosen to make the interviews as informal in nature as possible and to encourage openness from the respondents. The handwritten notes were transcribed on a computer shortly after each interview (usually directly after, and in all cases within 24 h) to minimize the risk that they would be difficult to fully understand and interpret afterwards. In two cases (interviews 10 and 12), a recording equipment was also used, since there was a risk entailed in those interviews to lose some of the information as the transcription had to wait for a few days. The result was thus one computer file from each interview, and each file contained a list of statements collected from that interview. In a few cases, the respondent drew pictures on a whiteboard, and these drawings were also captured in the notes. In some cases also presentation material was briefly used by the respondent in order to explain some things, and copies of these presentations were collected by the researchers. As shown in Table 1 above, the duration of the interviews varied between 45 and 120 min, with an average of 67 min.

Analysis framework

The processed result of each interview was a list of statements, in a Microsoft Office Excel spreadsheet, and each statement was tagged with a unique identity indicating the interview it originated from and the sequence number in that interview, to allow tracing back to the source. Some of the statements contained key findings that could bring light to the RQs, whereas others were more contextual, e.g., gave concrete examples of a specific situation in a certain company, but did not directly address the topics of interest. The latter were categorized as background information, and were removed from the structured analysis. The classification of which statements constituted key findings was done in the spreadsheet by two researchers independently, and then their analyses were compared. The key findings were then reformulated to become more general (e.g., the names of companies and people, and other specific information, scenarios were removed), and simultaneously translated from Swedish to English. The reformulation was reviewed by two researchers.

The statements considered as key findings were further classified into a number of areas, using columns in the spreadsheet, and the categories used for this were the building blocks identified in the Business Model Canvas (BMC) [8]. The reason for basing the analysis of business models on the BMC was that it is an established method for business analysis with clearly structured categories, where several categories focus on relationships between actors (RQ1), which are of particular interest to this study. The method is designed for the purpose of analyzing new business models, which is also highly relevant and applicable in the case of FES ecosystems (RQ2a). Also, after the BMC creation, the extraction of the product architectures, processes, methods and tools (RQ2b-c) was carried out as an additional step to complete this study.

The categories of the BMC are illustrated in Fig. 2, and are further described next:

- *Customer Segments* define the different groups of people or organizations an enterprise aims to reach and serve.
- *Value Proposition* identifies the products and services that create value for a specific customer segment.
- *Channels* describe how a company communicates with and reaches its customer segments to deliver a value proposition.
- *Customer Relationships* define the types of relationships a company establishes with specific customer segments.
- *Revenue Streams* represent the cash a company generates from each customer segment.

- *Key Resources* describe the most important assets required to make a business model work.
- *Key Activities* are the most important things a company must do to make its business model work.
- *Key Partners* describe the network of suppliers and partners that make the business model work.
- *Cost Structure* describes all costs incurred to operate a business model.

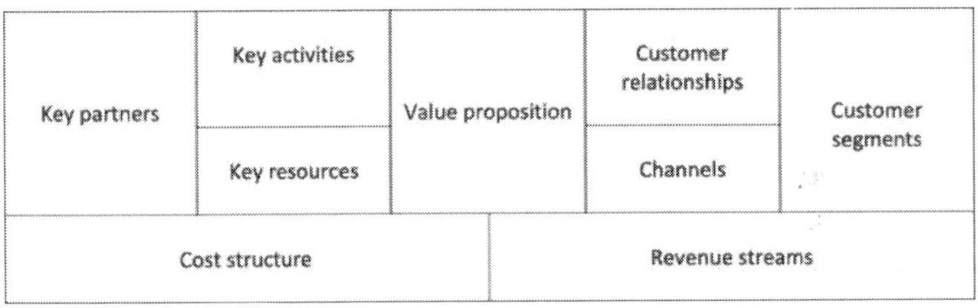

Figure 2. The Business Model Canvas (BMC).

By analyzing the interview data based on the BMC categories, it becomes possible to also extract elements to answer the different RQs. In particular, the actors and their relations (RQ1) can be found primarily by studying the Customer Segment and Key Partners categories; the product architecture (RQ2b) appears primarily as a Key Resource; and the processes, methods, and tools (RQ2c) are found as Key Resources and Key Activities. Information on the business model (RQ2a) are found in most categories, and especially in Value Proposition, Revenue Streams, and Cost Structure.

In our analysis, we chose to apply the BMC for the ecosystem as a whole, whereas a more refined analysis could have included one canvas for each actor type, or even for each individual company. Each interview statement was classified as belonging to several of the BMC categories. The classification was performed initially by the researcher that conducted each interview, and then a second researcher repeated the classification activity independently. In the end, all key findings that at least one of the researchers assigned to be included in the analysis of that category was used.

To illustrate the process used in the analysis, a concrete and representative example will now be provided. In one of the interview transcripts, the following statement was included (shown in an English translation, although in reality it was in Swedish, and with the information that could reveal the company made anonymous):

> *"[Name of company] delivers all [type of product] with a telematics unit, and contracts have been made so that the cost for data traffic is included in the service agreement."*

The statement was reviewed by two researchers, who agreed that this had sufficient relevant information in it to qualify as a key finding. However, it contained information about the company that had to be made anonymous, and therefore it was reformulated, while at the same time also being translated to English, resulting in:

> *"Telematics units could be included in the base product, and the data traffic fees could be included in a service agreement."*

Then, one researcher made the classification based on BMC, and decided that this key finding related to the categories Revenue Streams (since it deals with service agreements) and Cost Structure (since it deals with data traffic fees). A second researcher reviewed the first classification, and proposed that the key finding also related to Value Proposition (since the service agreement is a value proposition) and Key Resources (since the telematics unit is part of the hardware architecture, which is a key resource). The two researchers discussed their views, and agreed that this key finding should be classified in all four categories.

The same process was followed for all interviews and statements. The 12 interviews conducted resulted in 370 statements, of which 231 were considered to be key findings (62%). On an average, each key finding was assigned to 2.9 categories from the BMC.Table 2 shows how the various key findings were assigned to each of the categories.

Table 2. Number of key findings assigned to each category.

Category	No. key findings
Customer Segments	21
Value Proposition	106
Channels	36
Customer Relationships	82
Revenue Streams	67
Key Resources	100
Key Activities	110
Key Partners	86
Cost Structure	52
Total number of category assignments	660

After classifying all key findings to categories, the statements assigned to each category were analyzed separately, by grouping similar items into related topics for which a common conclusion could be drawn. These topics are discussed in the next section.

ANALYSIS

We now present the analysis of the interview data based on how the different aspects of BMC are reflected in ecosystems for FES, as identified in our interviews. Firstly, we provide a section with a short description of the key actors that have been identified from the interviews to address RQ1, and this is because the actors appear frequently in the discussion of the remaining analyses.

In the text below, quotations are sometimes used. Note that these refer to the generalized descriptions of key findings formulated in Step 4 of Fig. 1, and are not direct quotations from the interviews, but still reflect a direct link to the raw data.

Actors

Based on the interview data, we have identified a number of different actors that play important roles in the ecosystem for FES:

- *Manufacturer:* an organization that manufactures and sells products which are ready for final use. The manufacturer is responsible for integrating products and their components.
- *Supplier:* an organization that provides components to a manufacturer (or to another supplier) based on their requirements.
- *Content provider:* an organization that provides data which can be used in the operation of a product.
- *Communication provider:* an organization that manages and maintains the communication networks between actors and products.
- *Service operator:* an organization that operates different services related to the products, usually based on information obtained from the ES in those products.
- *Add-on developer:* an organization or individual that develops plug-in software that can be integrated into a product after production.
- *Owner:* an organization or individual that purchases the product offered by the manufacturer and the services offered by the service operator.
- *End user:* an individual who operates the final product.
- *Regulatory agency:* a public authority that puts restrictions on the technical solutions and operation of the products.
- *Information broker:* an organization that trades information, such as operational data from ES.

The above actors are role descriptions, and in a real ecosystem, the same organization could be playing several roles. Comparing these actors with the abstract categories identified by Manikas and Hansen [9], the manufacturer is usually the "orchestrator" and the "vendor", and the suppliers are "niche players". Content providers, service operators, add-on developers, and information brokers are all "external actors", and the owners and end users are "customers".

Within many organizations, such as manufacturers and suppliers, there are also important internal actors, e.g., engineering and production. For example one of the subjects described that the engineering department could be responsible for developing the product of the company, tailoring it to specific customer needs, and for providing advanced consultancy services.

Customer Segments

The Customer Segments building block of the BMC includes the identification of the most significant customer for whom the products create value. In the actors description in the previous section, the customers are primarily represented by the Owner and End user actors. To a large extent, the concrete Customer Segments are specific to the applications, and the interview subjects also did not discuss much about customer segments (as can be seen in Table 2). However, some general observations were made from the interviews, relating to broadening customer reach, improving customer feedback, and offering customer differentiation.

Broader customer reach

The customers are individuals or organizations that buy the physical products where the ES is integrated. The possibility of having add-on software for these products and developing new services, introduce new features and extended functionality in the products. Thus, as the products are extended, the market is also extended to include niche segments that were previously ignored since they were considered unprofitable. The main reason why customer reach is currently not developed to the full extent for services, according to one participant of the interviews, is: "Manufacturers are good at mapping products to different customer segments, but not so good at doing the mapping for services. It is not always the case that the same customer segments are relevant for services as for products." It would be beneficial to find ways to facilitate for manufacturers to identify and target different customer segments, based on the products, as well as based on the services offering something which is not currently successfully performed to the extent it could be.

Improved customer feedback

Manufacturers require market and end user feedback in different ways to better understand the customer segments. As one of the subjects mentioned in the interview, the engineering department is responsible for developing the product as well as for tailoring it to specific customer needs. However, planned new product generations are troublesome as the engineering department often lacks the ability to efficiently provide appropriate configurations and differentiated products. "It is interesting to get feedback from the field regarding how products are being used, in particular by early adopters." Thus, building a close relation between the development organization, collaborating parties and customer segments gives opportunities for obtaining immediate feedback from the field. Such feedback is significant, especially if it comes from early adopters of new products and services, since it can be expected to eventually reduce lead time and increase quality as well as provide an improved understanding of the relevant customer segments and their needs. However, feedback poses another challenge, related to privacy. Specifically, the information that is obtained from the use of products in various contexts and by various types of customers, needs to be protected, so sensitive information is secured and anonymous. Thus, finding ways to enable privacy, trust and confidentiality in the communications between the parties is a fundamental challenge. We will return to this issue in Section 3.5.3 below.

Customer differentiation

A platform that is flexible and provides the means for feature adaptation could be proven particularly useful when trying to reach new and diversified customer segments for services and products. More flexible architectures will be overall more easily adapted to changes, even after initial deployment, which creates new opportunities for all actors.

As mentioned in the interviews, organizations must differentiate their product and service offerings to satisfy different actors and customer segments. A platform that supports adaptability at different levels in different parts is a prerequisite for that to happen. In order to extend the market, the key customer segments need to be identified and collaborations and partnerships must be established. This can result in the definition of new products based on the market needs. Moreover, new features will be tested and evolved based on the observed markets' reaction.

Value Proposition

The Value Proposition building block of the BMC describes the bundle of products and services that create value for a specific Customer Segment. In this section, the focus will be on defining different service types, discussing the value of openness, and what are the most important quality attributes for FES.

Service types

A number of examples of products and services related to FES have been provided during the interviews, and based on our analysis, we can broadly classify them into the following categories:

- *Product services:* that improve the direct operation of a product in isolation. An example could be a connected navigation system in a car that not only shows maps but also traffic information. These services are often directed towards the end user of the product.
- *Process services:* that improve the operation of a product as part of a larger process (i.e., as part of a SoS). As an example, a truck fleet operator could gather data from all vehicles in a central system and improve navigation directions to drivers. These are often oriented towards the product owners, and mainly fetch data, with only limited control of individual products.
- *Life cycle services:* that use data from products to improve the associated life cycle processes. This includes predictive and preventive maintenance, but also feedback to product development on the usage of the individual products. These services are often directed towards the manufacturer or associated organizations (see Section3.2.2 above).
- *Extended services:* that use data from a product to improve the operation of totally unrelated products or services. For instance, vehicles could collect data about the status of the roads (e.g.,

detecting pot holes), and this data could be used by road administrators to improve road maintenance. They could also supply such data to an information broker, who would then sell it to other actors in the FES.

Open interfaces
Open interfaces, once in place, provide a value on their own for products, since they would make it feasible to integrate products with other products, forming a SoS. However, the add-on software would enhance these interfaces considerably, since it would allow much more extensive tailoring than, for example, an open serial bus interface. For process services, it is a common practice that products from different manufacturers are used within the same process. Hence, a product that is adaptable to, and is efficient in, such a SoS would be of value and particularly desirable to the owner. There are however, two contradictory trends: on the one hand, process operators wish more support in creating a complete system solution, and on the other hand, they strive to cut costs by purchasing equipment from competing manufacturers. Finally, being able to offer open interfaces of ES to IT systems, e.g., for business administration, could also improve considerably the owner's overall business processes.

Quality attributes
It was discussed during the interviews that having a vast amount of different add-on software for a product is important for ES, but not as critical as it is for smart phones. Nevertheless, it is crucial to offer a clear value to the specific products. Quality attributes, such as a usability, reliability and safety, were discussed and are expected to be important for ES, and assign specific restrictions on just how open the development could possibly get. The information stored in the system can also have different characteristics, based on the defined quality of the data itself, the acceptable delays, and so on. These characteristics depend on how the information will be used by a specific function and in different contexts.

Channels

The Channels building block in the BMC describes how a company communicates with and reaches to its Customer Segments to deliver a Value Proposition. In relation to FES, the interview subjects discuss four types of channels, namely sales channels, distribution channels, configuration channels, and information channels. They also highlight the importance of who is controlling these channels. The main difference with the FES setup is that manufacturers will be able to handle sales, distribution, and configuration directly, reaching their customer anywhere in the world over the Internet, and thereby reducing the need for certain special service providers (such as workshops) that exist today.

Sales channel

In traditional ES, the sales channel for software has been the same as for the overall product, and software has been delivered initially as pre-loaded into that system. Any configuration or distribution of upgrades has been required to be handled by service personnel, e.g. at a service station. The online customers' reach, through the Internet, anytime, from and to anywhere in the world, allows the possibility to manufacturers to have direct contact with the customer, with the potential of conducting additional sales of their products and services. However, manufacturers currently find it hard to quantify the value of this relationship in their business cases for connectivity, probably due to a lack of evidence concerning the magnitude of the additional sales.

Distribution channel

The distribution channel ensures that add-on software that has been made available to a certain system instance gets installed into that system. Much of the functionality of this channel is probably generic, and hence some of the manufacturers mention the possibility to outsource the operation of this communication channel to a third party, e.g., a telecom provider. It should be noted that there is also a need expressed of a distribution channel for testing purposes, where add-on developers can make their prototype software available to specific users.

Configuration channel

The configuration channel gives the possibility for the users to configure what kind of add-on software they have installed in their system, and modify it, either manually from a distance, through a desktop computer, or automatically, update to the latest version. By keeping track of what software customers have installed in their system, the manufacturer can also tailor their offerings to increase overall sales.

Information channel

Although the focus of the discussions in the interviews was on channels for add-on software, it is also recognized that similar channels are needed for information exchange. Hence, there could be a need for an information broker, whose main operation is to sell basic data to different service providers. Access to data over a public network would also provide a possibility for tuning and for individualizing equipment maintenance.

Channel ownership

In many practical situations, the above channels could all be implemented by the same IT system, but other possibilities exist. If channel responsibilities are allocated to several IT systems, channel ownership and control become critical factors, and to many, it seems natural that the manufacturer is responsible for the channels of their products. However, for add-on developers who provide software to products from many manufacturers, it might be desirable to provide their own channel, in particular if the end users purchase products of different brands and want to include the same plug-ins in all of them. There is also a fear that the manufacturers would take a large part of the revenue if all sales go through their channel (as can be seen in some channels used for apps in smart phones), and that an add-on developer's brand would be less visible.

Customer Relationships

The connectivity provided in FES gives the actors new possibilities to build relationships with their customers. By tying the ES into the business IT infrastructure systems of the customer, it becomes more difficult later on to switch to the product of a competitor. However, the technology also raises issues related to handling the services over the product life cycle, liability concerns, and dealing with the ownership of information, including the related privacy concerns.

Product life cycle

Many industrial products, that may be the target of FES, have very long lifetime, whereas the add-on developers and service operators can be expected to often be smaller and younger companies. Most likely, many customers will turn to the manufacturers to provide guarantees of the services associated with their products. However, in our interviews, there are clear indications that many manufacturers intend to make no commitment towards what services will be provided over time, and if such commitments should be made, there are no specific guidelines for what compensation should be offered, in case the service nevertheless ceases to exist. For many products, it is quite common that they also are re-sold second hand, and this leads to further issues related to the ownership of the installed add-ons, and service commitments. The considerations include whether they should be automatically transferred to the new owner, or if there are other possibilities that should better take place.

Liability

In general, product liability becomes complex when working in an ecosystem where different actors can add functionality. Some would claim that the manufacturer always has a liability for their product vis-à-vis the customer, but it is unclear how to handle this when third-party software is added, since it can often be hard to sort out if a problem was caused by the base product or the add-on. Most likely, the manufacturer will only take a limited responsibility for the add-on software. Many ES also have to meet legal requirements regarding safety, and are subject to certification by regulatory agencies. Today, in the FES domain these requirements do not often foresee the existence of add-on software.

Ownership of information and privacy
Networked systems like FES have enormous possibilities for collecting data, and this information is valuable to many actors. Therefore, some manufacturers and service operators require the users to grant them the right to access certain data in the products for other purposes than to provide a service to the same customer. In our interviews, contradictory opinions were voiced regarding if this constitutes a problem with respect to the privacy of the users. Some respondents see this as a major problem that has led to vivid debate within their companies, whereas others claim that as long as the identity of individuals is kept confidential it is not a problem. Possibly, the difference in perspective of each respondent depends on if the product is consumer- or business-oriented, where privacy of individuals is a larger concern for consumers and less so for companies, where employees expect to be subject to a certain degree of surveillance. Also, it is noted that the importance of privacy is to some extent situation dependent, e.g., in an emergency situation the location of a missing person might be very important to share, whereas under normal circumstances it should be kept confidential.

Revenue Streams
The Revenue Streams building block of the BMC represents the cash a company generates from each Customer Segment. In this section, different ways of generating revenue are discussed. From the interviews, it is apparent that "there are not any established models for paying for services and data." It is unlikely to see one model to fit all, but different models will most likely be needed in different situations.

Volume increases

In current ES-based industries, there is no specific revenue stream defined for software, and the prospect of FES could thereby improve the revenue for manufacturers in several ways. One way is to make their products more attractive with the added functionality and increased flexibility, thus increasing earnings from goods through increasing the volume of the offering, e.g., by reaching niche segments as discussed in Section 3.2.1 above.

Recurring sales

FES also offers possibilities for recurring sales of services during the lifetime of the product, and not obtaining profit just from the initial selling of the product. In many industries, as products mature they are commoditized, and the manufacturers need to offer new values, such as overall business and system solutions and services, to maintain their profitability.

Direct sales of add-on software

One reason for manufacturers to open up their products to third-party software suppliers in FES is that they can improve innovation processes that can provide an extended customer value. The add-on developers can receive revenues through either direct sales of their software to users, or by developing the add-ons based on a contract with the manufacturer or owner. Whichever way is used, it is important to find financial solutions that give a sustainable profit to both sides. Different pricing schemes are possible, including offerings of a complete package with all available add-ons, without any additional usage cost; providing an entry-level of functionality for free, and extended features at a premium price; providing trial-versions for a limited time or with limited functionality, and then requiring payment to continue using the offering afterwards.

Subscription fees

Several interviews mention the possibility of using subscription fees as an alternative to selling an asset. "Subscription could be an interesting business model. However, manufacturers do not have the experience of handling this." Some consider it would be better to give unlimited access to information services based on an overall service agreement rather than having to deal with a costly management of small revenue streams from many singular services. Anyhow, it is important to find ways for getting paid for information, to ensure that the services can be maintained over time and hence allow users to build confidence for the subscription scheme.

Revenue sharing

"Revenue sharing is an interesting business model. For instance, if the service promises to achieve x% efficiency improvement, the profit the customer makes from this improvement could be shared with the actor offering the service." Some information service operators are also interested in revenue-sharing models with manufacturers, since having their service available in the manufacturer's product gives them another channel to the customer, which could motivate sharing the revenues with the manufacturer.

Key Resources

The subjects in the study span the range from enterprises with worldwide operations, to small and medium enterprises with niche products and small organizations. The main resources used by the partners of the ecosystem involve the products, services and their add-on components; key people involved in the principal activities; the platforms used for the requirements definition, development, integration, testing and maintenance; the distribution channels as well as the assets owned by the organization taking the central role in the development of the products and services (i.e., the manufacturer).

Our study revealed other issues which are more related with the intellectual properties and human resources, such as, the hardware and software constructed within the ES domain and the know-how around them. However, in the envisioned scenario of FES that motivates our study, additional intellectual resources, such as, brands, patents, and standards are of key importance. The principal challenge is best illustrated with the following statement from one subject: "How should one be able to protect intellectual property of the add-on software that is installed?"

Development environment

Currently, the manufacturers in our study provide ES packages containing both hardware and software, where the two are closely linked to each other. Most manufacturers open up their products to some degree for their customers, providing opportunities for black-box configuration and programming. However, they rarely open up for third-party extensions, for example for add-ons. One of the problems they describe, may be illustrated by the statement: "an add-on software developer needs access to some kind of representative hardware for testing". In the IT domain, systems mainly comprise of software, which is easy and inexpensive to produce and transport. It is also easy to provide connectivity using, for instance, a network to connect to another remote system. However, for ES the custom hardware and even mechanical parts add some complexity and are also important when it comes to considering development, integration, testing and maintenance. In particular, it is required for testing by a third party, and is expensive to reproduce and transport. This is considered an impediment for opening up to small niche players, as the up-front investments are quite high, especially when considering scenarios where they would like to provide integration of products from several manufacturers.

Human capital

Several companies in the case study include services in their value propositions. This implies that human resources and assets, such as the experience and know-how, are key resources for a company. A company's collective experience from deploying their products in a specific domain is transferred to clients. From the company perspective, their solution, including products and services, is applied on a particular problem. Opening up products would pave the way for niche players that approach the problem from the opposite perspective, i.e., from the problem domain. They can integrate and combine hardware and software from several players targeting an optimal solution for a specific niche problem. This will also increase the market competition, so that all players would benefit from it in the long-term.

However, open ecosystems may pose threats to players having as asset highly skilled staff. It would create several opportunities for key employees to set up their own startups or leave the ecosystem for collaborating with other external parties. The consequence is a flight of human resources (brain drain) that may constitute a severe risk to a company. Another risk is the flexibility induced by the openness that may cause quicker and more frequent shifts in market trends. As highlighted in one of the interviews "it is a challenge to try new business models and adapt the organization". The players must invest to keep up with the pace if new products and add-ons appear more frequently on the market. Activities related to knowledge management and training become key for the ecosystem actors.

Branding

Another important issue regarding the key resources is branding. Consider a situation today where one company provides both hardware and software. The company uses displays and possibly the hardware for branding, for instance, providing a company specific user interface and logotypes. This causes several challenging problems in an open-market scenario. As expressed by one of the niche players in the study: "would it be allowed to show our logotype in the display of the product"? In some situations the answer is no, in others it is yes. Sometimes, "if a third-party developer has a very strong brand, they could be allowed to show their logotype in the display of a manufacturer's product".

Standards

Things like branding and patents are important to protect resources of a certain actor, but an alternative route is to standardize different parts of the system instead. This can be either carried out through established standardization bodies, or using de facto industry standards based on one manufacturer's (or a group of actors') solutions that are adopted by other actors. The need for standards is expressed by representatives of many actors in our interviews, including manufacturers, suppliers, add-on suppliers, and also indirectly through customers. "Customers are complaining that they need to invest in expensive add-on equipment with uncertain value, and they are pushing for more standards." However, it is far from trivial to develop standards, in particular in an area which is emerging and has not yet matured. The innovation potential in FES is high, but standards can easily become conservative and restrictive, and the process of developing a standard can take very long time.

A conclusive statement from the analysis regarding key resources is that the challenges lie in finding a balance between openness and protecting intellectual properties. The overall value of the key resources may be increased if they are integrated with other resources. However, the common viewpoint is that the associated risks are more prominent and difficult to deal with.

Key Activities

Key activities in the BMC capture the most important things a company does to make its business models work. The main activities that need to be carried out include defining the product and its architecture, and developing, integrating, delivering and maintaining the product and the services around it. With respect to the business model other activities include building a stable collaboration network, offering services such as consulting, data trafficking, security, marketing, sales, and supply network support. In this section, we will focus on production related activities, and activities for maintaining product line components to preserve their competiveness.

Production

The manufacturers in the study typically own the product from the phase of conception until production. They design a product and manage the production and target to optimize both accordingly. In the envisioned scenario of FES, hardware and software may be offered by third parties, which could cause some implications, as raised in the interviews. However, "add-on equipment with physical components could be fitted on the production line of the manufacturer, or at a separate company". In the more mature domains in this study, like the automotive, that is often common practice. Vehicles are equipped during production or before delivery. For example a truck-mounted crane is often ordered together with the truck, delivered by a third party, and installed by the vehicle manufacturer during production. This implies that the product architecture and production activities are prepared for the add-ons. One such example is when a truck provides an interface that an add-on may utilize to enable the truck driver to operate the add-on from the truck cab.

The manufacturers in the study represent a mature domain with respect to production. Activities for integrating hardware add-ons on key resources are commonplace practice. However, opening up the activities in respect to software resources is considered challenging. Companies are not using software to the same extent and have instead relying more on electro-mechanical and hydraulic interfaces. Software integration is new to these companies and thus it poses several challenging activities that they have not yet addressed.

Product line management

The interviews depict a current landscape where key resources are proprietary and manufacturers have full control over all aspects, including management and evolution. "The market department almost always defines new products and features based on customer contacts." When forming ecosystems for FES, activities related to evolution require more extensive coordination among the actors. Changing or removing Application Programming Interfaces (API) and similar activities may have dramatic effects on other parties in the ecosystem and could even result in niche players being forced off the market. This is the technical perspective on the discussion included in Section 3.5.1. However, open ecosystems would introduce a possibility for manufacturers to reduce risks concerned with evolution, given that third parties will take on that responsibility to some extent. This will reduce evolution-related costs and could have an indirect effect on revenues.

A question raised in the interviews is how open the evolution will be. "How difficult will it be to get permission from the manufacturer to add software?" This implies that some manufacturers hold a reluctant position that may possibly hinder a flourishing ecosystem. It is clear that management and evolution activities, to some extent, must be opened up and should be appropriately coordinated among the actors. As a consequence, actors may create new value propositions that generate new revenue streams, for instance, developer licenses that provide guarantees with respect to support.

A recurring concern in the interviews is the issue of assurances. Verification and validation (V&V) within the process are considered key activities. The interviews have described several scenarios and models for this, ranging from offering V&V as a service to third parties, and third parties performing V&V themselves. It is clear that this area requires a more in-depth analysis, given the complexity of the activities and the degree of V&V criticality for product lines.

Manufacturers and third parties also perform brand-related activities, including market analysis, campaigns and rebranding. In mature domains, these activities are already well-known and well-coordinated, with respect to hardware. However, when opening up products and markets to other key actors and niche players that enter the ecosystem, the landscape is bound to change dramatically, and thus, all of these activities will need some degree of support and re-definition.

Key Partners

The Key Partners building block of the BMC represents the network of actors that make the business model work. The key partners in the ES ecosystem have already been mentioned in Section 3.1 above. In this section, we will mostly focus on their relations and issues associated with this.

Traditional supply chain

Just as for traditional ES, FES will also use a basic supply chain where the primary actor is the *manufacturer*, responsible to manufacture and sell products. Often, manufacturers are also responsible for the product integration. The main partnerships of the manufacturer are formed with *suppliers* providing components to be integrated in products, and with *content providers* providing data to different actors. The content provider may be the one responsible for also distributing the data, or might have another partner to provide the required connectivity (e.g., a telecommunications operator), for transmitting the data. In general, the core partner network composed from these four partner types (manufacturer, supplier, content provider and communications provider) is typically controlled by the manufacturer.

New actors

In a more open business model, new actors appear and the traditional supply chain is not sufficient, but further partnerships can be formed dynamically. The analysis of case studies from large enterprises is an emerging need, while also transferring lessons to small and medium enterprises is a challenge in the field. The operation details of such a business model are complicated, because it expands all current operations of actors and also new actors are entering in the model. Adapting the organization to becoming able to receive them is challenging for several reasons. First of all, the span of the model, human capital and operational management difficulty increases. This brings in challenges of who is in control and what are the architectural and infrastructural needs to support the business model. Secondly, accumulating efforts to direct services and their revenues, especially when they are entailing many international or transnational organizations can be quite complicated.

In this model, new services are developed and also new actors emerge. One new partner is the *service operator* with the ability to operate new services related to the products. The products might even not be originally offered by the base product, but emerge or are extended later on, with new features. In addition, other new actors are *add-on developers*who develop software and hardware that can be integrated into the base product. They are typically younger and smaller organizations than manufacturers or suppliers, and are concentrated around ideas to satisfy a new market segment, create a new innovative product or component, create a better value, extend collaborations, offer a new service, and utilize latest technologies.

Add-on developers play a secondary role in the ES ecosystem today; however, they have a prospect of rapid growth by attracting new customers and partners. These organizations face the following challenges: identifying what is not currently offered by large manufacturers; deciding where to invest (e.g., in products, components or services with relatively low profitability for the manufacturers, where the competition and risk is lower); and facing high cost barriers for entering the ecosystem. Manufacturers that participated in our study mentioned that sometimes they decide to leave a market segment to other players. This is usually because the cost of maintaining expertise in a particular domain may be too high compared to the revenue it creates. The manufacturer thus sometimes instead focuses on supporting players that could act in that segment. They trade the risks related to market presence for a reduced share of the revenue stream.

Cost Structure

The Cost Structure component of the BMC describes all costs incurred to operate a business model. In the interviews, four general groups of costs were discussed, namely development costs, product costs, operating costs, and information costs.

Development cost

For manufacturers, the development costs do not appear to be radically different for FES compared to traditional ES, since it is basically a matter of providing some additional capabilities in the platform, which is anyway something carried out once. However, for add-on developers, development cost is a larger issue. In many situations, add-on developers see the need to adapt the same add-on to different platforms of several manufacturers in order to amortize cost on a larger user base. "The effort to adapt a solution to the base system of different manufacturers is as large as developing the functionality." In particular, this is an issue when there is a Human Machine Interface (HMI) involved. It also leads to an enormous problem with variant handling, and evidence of this can already be seen in the smart phone industry, where certain apps exist in many hundred variants for different phones. Therefore, it is essential to provide reuse possibilities for the software components that are used to build add-ons. In addition, there might be sometimes a need for rather expensive hardware to stand as a representative test environment (see also Section 3.7.1), and therefore, manufacturers need to consider how they can better provide shared facilities for testing purposes in the ecosystem. On the other hand, by having a shared infrastructure for the different channels, economies of scales in the server side systems outside the physical products can be achieved.

Product cost

Even though manufacturers are not so concerned about the development costs for FES, they are the more worried about the product costs. Flexible solutions, like add-on software, will certainly require surplus capability in the hardware (e.g., processing power, memory, input/output), especially if graphic displays are involved since they require large volumes of data. Many of the manufacturers are traditionally very cost-focused organizations, and they find it difficult to discuss adding this flexibility in their business cases. This may considerably slow down their transformation from product-centric towards service-centric companies. For the user, on the other hand, physical add-on equipment is usually much more expensive than an integrated software solution that uses sensors and actuators already present in the product.

Operating cost

The most evident operating cost is the cost required for data traffic, which is in particular a concern for mobile ES that need to rely on wireless telecom services, sometimes involving coverage costs and expensive roaming when traveling to different countries. Different models for sharing these costs between the actors are discussed in the interviews. One possibility is to have all costs included in some higher-level service agreement between the manufacturer and the owner. Another possibility is to have users pay for data traffic directly related to the user features, whereas the manufacturer pays for traffic related to services, e.g., diagnostics and field data follow-up. A third possibility is that the owner covers all costs directly by providing their own telecom subscription. There is also a direct relationship with the operating costs and the quality of service, since high-quality services will require more frequent and accurate data updates, leading to higher bandwidth consumption.

Information cost

Another category of costs discussed in the interviews is the information costs. In the future, users will probably expect something in return for letting anyone take advantage of data originating from the systems they operate, either monetary revenue or new or improved service offerings. This could result in some cost for example to a manufacturer who is using such data for obtaining feedback about their products and services in the field, or an independent service operator who is aggregating data from many users to provide a new service.

Summary

To summarize the analysis using the BMC, the key elements of each category are shown in Fig. 3.

<table>
<tr>
<td rowspan="2">Key partners
• Manufacturer
• Supplier
• Content provider
• Communication provider
• Service operator
• Add-on developer
• Owner
• End user
• Regulatory agency
• Information broker</td>
<td>Key activities
• Production
• Product line management</td>
<td rowspan="2">Value proposition
• Service types
o Product
o Process
o Life cycle
o Extended
• Open interfaces
• Quality attributes</td>
<td>Customer relationships
• Product life cycle
• Liability
• Ownership of information and privacy</td>
<td rowspan="2">Customer segments
• Broader customer reach
• Improved customer feedback
• Customer differentiation</td>
</tr>
<tr>
<td>Key resources
• Development environment
• Human capital
• Branding
• Standards</td>
<td>Channels
• Sales
• Distribution
• Configuration
• Information
• Channel ownership</td>
</tr>
<tr>
<td colspan="2">Cost structure
• Development cost
• Product cost
• Operating cost
• Information cost</td>
<td colspan="3">Revenue streams
• Volume increases
• Recurring sales
• Direct sales of software
• Subscription fees
• Revenue sharing</td>
</tr>
</table>

Figure 3. BMC summarizing the key elements of the analysis.

DISCUSSION

The analysis based on the BMC provides sufficient information to address our business-oriented RQs, and especially the RQ1 and RQ2a related to actors and business models. In our research, we are also interested in what are the particular implications of FES on architectures, processes, methods, and tools, as indicated in RQ2b-c. In this section, we revisit our RQs to discuss and summarize our findings and identify a number of critical challenges. We also discuss some overarching challenges that cut across the key elements of the ecosystem.

4.1. Ecosystem actors and relations

Our first RQ is concerned with the ecosystem's actors and their relationships, which constitute the business architecture backbone. Sections 3.1 and 3.9 discuss actors and to some extent their relationships. One observation from the interviews is that a traditional supply chain with manufacturers and suppliers will still be at place as the core. However, we have identified developments involving existing and new actors that are likely to extend this structure and introduce new challenges to the ecosystem. These actors are the add-on developers, the information brokers, and the regulatory agencies (presented in Section 3.1).

Challenge – Explicit and implicit relationships

The *manufacturers* and *suppliers* are key actors in the traditional supply-chain based ecosystem. In such ecosystems the relationships are well defined and explicit, and therefore relatively easy to trace and understand. For instance, when the suppliers and a manufacturer have agreed to a particular standard all parties are aware of that agreement. However, the introduction of add-on developers and information brokers in the ecosystem would unavoidably shift this balance. A typical scenario for an add-on developer is "providing a new product through integration of two or more existing products". In the scenario, interactions with other actors use well-established, explicit relationships. However, some explicit relationships may become implicit for the add-on developer, and as such, they are more difficult to identify and therefore they constitute a potential risk, if they are not properly managed. One such risk may occur when an actor integrates systems without direct involvement with the sub systems' original manufacturer or suppliers. The subsystems may for instance have dependencies on artifacts provided by a regulatory agency. These artifacts provide specific rules and regulations that may be violated by the integration if their results are not known by the integrating actor. The challenge is to make all critical relationships explicit for the actors in the ecosystem.

Challenge – Flexible authority

In the interviews, several example scenarios were described where the ecosystem's authority was granted to several actors, or even distributed across several actors, and this was changing over time. Authority in this case, refers to the form and responsibility of control for the ecosystem products and information artifacts, channels, etc. The add-on developer and information broker actors are two concrete examples that take over authority from manufacturers and suppliers; add-on developers take over control when they are integrating systems or adding functionality, and information brokers when they are aggregating and redistributing information. The challenges are among the developing models and protocols managing these phenomena. This requires a better understanding of the underlying rules that govern these processes and how the models should be instantiated for a specific ecosystem.

Business and technical architecture

With our second RQ we target additional key elements of the business (RQ2a) and technical (RQ2b) architectures and their relationships. The case study participants described their current architecture, and discussed issues and concerns from different points-of-view and at different levels of abstraction. A general observation is that the concept of architecture spans the range from high-level business architecture involving components, such as business models, to low-level components that represent a block in a control algorithm.

The analysis of the collected material identified a number of key elements as reported above in Section 3. We have also identified four key challenges related to elements in the architectures and their inter-relationships:

1. Business and technical architecture alignment, which is concerned with a comprehensive architecture that covers all aspects of the company's business and key resources.
2. Openness, which deals with the extensibility of products and processes, including management and the technical aspects involved.
3. Assurances are to a large extent cross-cutting the former categories. They are concerned with how product, service and process qualities are achieved and what type of guarantees regarding the respective qualities are derived and how.

4. Configuration is concerned with product instantiation and configuration that involves multiple stakeholders.

Challenge - Business and technical architecture alignment
Several case study participants expressed that the architecture was the key success factor for their business. However, their mapping of business to processes and products posed a great challenge to them. A direct quote from one subject captures the essence: "What should the architecture and infrastructure be like to support the business model?" This enterprise architecture challenge is well known and has been addressed in research and practice for many years now. However, in the envisioned landscape which introduces more openness and integration, add-on developers add to the complexity of the problem and impose required re-factorings of the enterprise architecture. Statements like "the complete offering only considers the company's products. No specific support is included for third-party products and services" illustrate this. The business and product architectures are not prepared for open business models.

Challenge - Openness
The challenges when moving towards more open processes and products that integrate with others were highlighted in several interviews. Subjects described it as: "Many ES are closed products under the full control of the manufacturer. How can they be opened up?" The subjects describe their current architectures as mature and the software architecture's component view is a direct mapping of the deployment architecture. The architectures define concrete subsystems and the interactions in-between subsystems are kept at a minimum. The primary cause is that it reduces the V&V effort.

Most subjects discussed openness at the technical architecture level. The subjects also talked about several risks and mitigation strategies in relation to this. When investigating the rationale for some of the risks we discovered that most were related to control and trust. A majority of the subjects are active in closed and regulated domains, where system behaviors and interactions are negotiated and stated in contracts or are regulated by standards and laws. In this landscape, it is understandable that actors are reluctant to open up, most likely due to the overhead it will introduce when the control over the complete supply-chain is decentralized and also the uncertainty with respect to conformance. Several subjects expressed concerns with respect to control and governance risks. However, they explained how they consider most of them as technical risks and described strategies for mitigating some of them. One strategy was to introduce a classification of architecture subsystems that explicitly expresses their degree of openness. This would reduce the technical issues, and as a consequence, also mitigate risks connected to trust and control. Several participants also advocated standards to regulate openness: "there is a need for industry standards when it comes to handling add-on software and other additions".

Challenge - Assurances

The concerns expressed with respect to openness are instantiations of the principal challenge in the domains where the case study subjects are active, namely safety. This challenge permeates the companies' architectures and explains the rationale for their closed processes and technical architectures. In these domains, owners and regulatory agencies require guarantees that the subjects' products behave according to specification. The manufacturers and add-on developers address this with determinism. Specific output as a result of specific input to processes, products, and their constituents must be predictable. However, opening up would introduce an element of non-determinism that reduces their ability to verify and ensure certain behavior, i.e., reducing or sometimes removing predictability.

One such example is mixed criticality, where some sub-system should behave according to one profile while another profile applies to other sub-systems. One subject stated, "real-time properties are also important but more easily managed". They mitigate risks related to real-time constraints by local control. This reduces the problem to provide assurances. However, if systems open up, then that same component may be used in another context, according to another profile where it is controlled externally. This introduces uncertainties related to communication, and predicting real-time behavior becomes a much harder problem since the assurances provided are no longer valid. The subjects discussed some strategies to mitigate these risks. Most of them were stating the same principles, i.e., more standards, more negotiation and more governance. One statement illustrates the subjects' views on this: "software for embedded applications requires safety, and can never be as open as it is for smart phones."

Challenge - Configuration

Several subjects describe that their products allow end-user configuration, including integration with add-on developer products and programming or configuration of specific components. A more open setting would provide end users and integrators with even more powerful means. For instance, the possibility to develop add-ons would improve the current situation considerably by introducing more flexibility. It would "be useful to adapt their system to the particular context and be able to communicate with the other systems in that context". This illustrates that manufacturers and add-on developers would benefit from more open architectures and improved configuration support, as it will improve the products' extensibility and flexibility.

Processes, methods, and tools

The focus of RQ2c is processes, methods, and tools, which form the infrastructure for how products are managed during the entire product life cycle. It captures practices, guidelines, and standards in an accessible format that is used to coordinate and perform work within an organization. In a federation of ES, coordination of processes poses a great challenge. Coordination is required in-between different actors and the actors' processes. The possibility to add hardware and software to a product increases the complexity of the product's life cycle as "there are [at least] three cycle times to handle: One for the physical product, one for the software platform, and one for add-ons [...]". In an ecosystem, we will also have dependencies between different products and services, and their life cycles.

Challenge - Quality assurance

Connecting with other actors in the design and production processes will require processes to be integrated to a large extent. Such integration might be implicitly or explicitly defined and described, and address internal or external integration. In the latter case, process standards must play a prominent role. For example, "manufacturers claim they need to develop add-ons themselves to ensure quality and a get a consistent HMI". This illustrates a situation where manufacturers and add-on developer processes should be integrated and possibly governed by standards. The motivation for such integration is product quality assurance, which is critical in the domains under study.

V&V is more complicated in a FES when compared to pure software ecosystems. Products consist of a mix of the manufacturers' hardware and software in combination with add-on hardware and software from possibly several add-on developers. They also might have interactions with other external and internal systems, forming a SoS. The safety-critical nature of most applications is the key contributor to the problem. The Directive on Machinery issued by the European Union [10] is an example of how to regulate development and testing of safety-critical applications. This directive (or similar documents) forms a basis in most domains for governing integration of add-on developer's parts in products or add-ons. Some subjects expressed that "some sort of certification is needed for add-on software". There were also examples of how to integrate processes and certify add-on developers' parts. For example, "developers of [add-on] software are given an API, a test environment, and then the manufacturer certifies the software".

"An add-on software developer needs access to some kind of representative hardware for testing" and at least "a simulation environment is necessary for third-party developers to test their plug-in modules". The challenge of testing software on dedicated, possibly expensive hardware is addressed by simulation and emulation. However, several subjects describe a situation where the manufacturer may use hardware or test environments to control the market and decide if a market segment is open or closed.

Challenge – Updates

The second area that several subjects discussed is updates. Updates also require extensive coordination, especially with the V&V processes, and this must be reflected in processes, methods and tools. "There is a need for an updating mechanism for installed add-on software". However, this update mechanism is not just about distributing and installing software updates. It must also make sure that the software being deployed is verified and validated, possibly certified to function with the installed hardware and software, which may include a set of software add-ons that have been developed and added to the system after its initial deployment. In addition, it must deliver similar assurances regarding reliability, stability, and performance as the ES functions. This exemplifies the complex web of developing process interactions that is required due to various quality assurance standards to guarantee safety. Managing these interactions is a key, and risks are mitigated by standards. This is probably why "things that are standard tend to survive".

Ecosystem challenges

In the analysis and discussion we have presented a number of challenges, most of which relate to opportunities and problems of the introduction of FES. We will now summarize what we consider to be novel overarching challenges at the FES ecosystem level that need to be addressed both by research and practitioners.

Challenge – Ecosystem management

A thriving ecosystem will most likely not emerge by itself, it will require stimulus from key actors, either manufacturers that create ecosystems around their own products, or possibly industry consortia that agree on certain common rules and standards. In FES it may be more challenging to identify key actors and the role as key actor may shift, for instance, from manufacturer to supplier for one value proposition, or from manufacturer to information broker for another proposition. To achieve *sustainability* in the ecosystem, finding ways of sharing both revenue and cost is essential, and this includes cost for products, services, and add-on software, but also for data where the functioning markets need to be organized. Finding a stable *architecture* is also critical, and this is not limited to the technical architecture, but also adheres to the enterprise architectures and all actors involved. The *open interfaces* play a central role. However, openness that can lead to new functions and services must be balanced against the dependability of the base system. Good interfaces are necessary, but not sufficient, to stimulate *open innovation*, and other means also need to be installed, for instance, support channels, meeting places for the actor communities, etc.

There appears to be significant differences in focus between consumer-oriented products and business-to-business products. For the former, customer features are mostly important, whereas for the latter, ways of improving efficiency matter more. Possibly, this could affect how ecosystems are managed. Another critical force that has an indirect effect on how the ecosystem may be managed is the regulatory agency actor. This actor provides rules and regulations that will affect parts of or the entire ecosystem, how actors behave, their relationships and obligations to other actors, etc.

Challenge – Product development

Once the ecosystem is in place, support is needed to develop the products. The challenges we identified include several that stand out as unique for the FES domain. The product domains of FES typically impose more strict requirements on processes, methods and tools. These requirements are often regulated and controlled by an outside authority. This has an effect on several actors, for instance, add-on developers need to have a *development environment* for the products of a particular manufacturer, and ways must be found where such an environment of sufficient quality can be created without imposing heavy costs for the manufacturer and in compliance with adequate regulations. In many cases, add-on developers will make versions of their software to fit several manufacturers' products, and this means that strategies for managing product lines are needed, both by the manufacturer and by the add-on developer, to minimize the adaptation cost. This again poses novel challenges with respect to relationships to different regulatory agencies.

Another challenge that contains unique novel aspects is quality assurance. In FES ecosystems the responsibility for *V&V* is in some way shared by manufacturers and add-on developers, but the exact distribution of work between the two needs to be sorted out in each case. The same holds for the manufacturer-information broker relationship. V&V also include issues of *liability*, i.e., who is in the end to blame if something goes wrong. In a *mixed criticality* environment, where safety-critical applications share a platform with other software, this is particularly important and constitutes an even more complex challenge as it varies from one product to another.

Challenge - Ownership and control

The *channels* for add-ons and information will be central in the ecosystem, and the ownership and control of those channels will give power to the actors who possess them. An open environment will make it harder to protect the *intellectual properties* of different parties, and either mechanisms have to be in place for this, or the actors have to accept the fact that this risk will not cease to exist, and assume that the benefits from participating in the ecosystem outweighs this drawback, as many actors within the open source software domain have already chosen to do so. Finally, there will be sensitive information about individuals and companies flowing in these systems, and adequate protection of this information, in respect to the *privacy* of individuals, anonymity and confidentiality, needed to gain *trust* in FES.

Challenge - Support for open innovation

One long-term goal for our research is to investigate and understand how to best cater for open innovation in the ecosystem of FES. Several of the challenges described above include problems and opportunities that require further investigations in this respect, as also highlighted in open software innovation [11] and [12]. Open innovation requires openness at the business and technical architecture levels. The emergence of new open business models needs to be supported with new roles and patterns of collaborations to cater for innovation and transparency [11]. Software ecosystems, conceived as networked organizations, are considered to contribute in understanding this new community aspect, extending across many (traditional) organizations [11]. The notions of openness with the adjustment of the smallest amount of innovation are enough for offering the competitive advantage to organizations that can effectively harvest its benefit[12]. However, several impediments that are unique to the FES exist. These impediments are a consequence of the products' domains and the requirements imposed by these domains, for instance, safety requirements and certification of system behavior by external bodies. The principal challenge to establish provisioning for open innovation in FES ecosystems translates into finding the right balance of management, development, ownership, and control in the ecosystem. A balance that makes it feasible for new and existing actors to extend the ecosystem with new value propositions and establishing sustainable revenue streams while fulfilling domain-specific requirements.

VALIDITY

We will now discuss the validity of the findings presented in this paper. A number of validity threats exist in case studies, and following the categories of [5], the threats in this case study are reported below, including some more detailed threats based on [13].

Construct validity

The construct validity ensures that the construction of the study actually relates to the problems stated in the RQs, and that the chosen sources of information are relevant. A specific threat to construct validity is the use of unclear terminology, which is certainly an issue when dealing with emerging concepts like FES. To reduce this threat, we have used the established BMC framework for structuring the analysis of the interview data and also, we have made sure throughout the interviews that a common understanding of the meaning of the terms used existed. Another possible threat is that the respondents might have guessed what hypothesis the researchers had in mind, and adapted their answers accordingly, for instance by exaggerating their opinions in an attempt to try to influence the outcome of the study. We tried to reduce this threat by using open-ended questions in the interviews conducted and by not explicitly discussing the hypotheses with the interviewees.

There is also a risk that the analysis is influenced by the researcher's expectations, so that the researcher only sees what he or she already believes. To reduce this risk, we tried not to invent our own structures of classification, but instead use an established framework (the BMC) and use open-ended questions. Also, the statements and findings were cross-checked by at least two independent researchers. Finally, respondents could be hesitant to express their views if they could later be affected by the responses, and to handle this risk, the anonymity of both respondents and the participating companies was guaranteed.

None of the respondents had dependencies on any of the researchers or the research outcomes. However, as pointed out in Section 2.2 above, there were established relations between the researchers and interview subjects beforehand. This is thus a potential source of bias, but also a facilitator for finding the right people to talk to, offering the appropriate level of trust that allowed an open discussion to be carried out, and being able to interpret the results correctly due to an extensive contextual knowledge already in place. One interview subject even stated explicitly that "since I know you, I can speak freely, trusting that you will understand what is confidential information and keep that to yourself."

Internal and conclusion validity

Internal and conclusion validity concern the possibility to ensure that the actual conclusions drawn are true. In [5], it is stated that "internal validity is only a concern for causal (or explanatory) case studies". This case study is highly explorative, and hence less sensitive to this threat. However, there are still issues that are relevant to mention.

One issue is the selection of respondents, which was made to ensure that they have the right knowledge about the context of the research and the challenges around it. By relying on already established contacts and having some prior knowledge about these individuals, this risk was minimized, and the long experience in the business area and in the specific company of all participants provided the possibility for gathering a rich variety of data.

The sample size of the individuals involved was relatively small, and does not allow for any deep quantitative analysis, but with a stratified selection, we ensured that many of the actors in the ecosystem were covered. The mortality was reasonable, with two persons opting not to participate in the interviews due to time restrictions.

Another risk is related to "fishing", i.e., that the researchers consciously or unconsciously search for certain answers. We tried to avoid this by having as open questions as possible in the interviews, and by finishing each interview by asking if the respondent felt that there was anything else that should be brought up and discussed.

External validity

External validity concerns how the results can be generalized. This is a specific concern for a case study, where it can always be discussed to what extent the observations are particular to a certain environment, or whether they are examples of general phenomena. The primary type of external validity is whether the conclusions can be generalized to a different organization, either within the same or in a different industry. In our work, we did include representatives from several business sectors and companies and we believe to have captured many characteristics that are general to the problem domain. Still, we cannot with certainty say that this is the case, and to enable us to draw such conclusions, further studies are needed.

Reliability

Reliability relates to the ability of others to replicate the study and arrive at the same results. A basis for replication is to have a well-documented study design and well-structured data collection processes, and we believe that this is the case for the study presented here. Assuming that the study was replicated and resulted in roughly the same transcripts of the interviews, it would still not guarantee that the resulting issues would be the same. There are different ways of interpreting the textual material, and in some cases there could be several ways of grouping statements resulting in different sets of abstractions. To reduce this risk, we had two researchers doing the classification of statements collected independently.

A particular threat to validity in this study is how the data was collected during the interviews. In 10 out of 12 interviews, handwritten notes were used, and in the remaining two, a recording equipment was used. There are pros and cons of both approaches. Clearly, there is a risk of missing important information when not recording interviews. This risk was reduced in the non-recorded interviews by transcribing the handwritten notes directly after the meetings, and one of the reasons for actually recording the remaining two interviews was to handle the risk that the period between interview and transcription could become a little longer in those two cases. However, a drawback of recording equipment is that some respondents may become hesitant to speak freely. In general, in this study, the topic is very central to the future business strategies of the companies, and thus the risk was that some information would be held back and thus reducing the value of the whole study (as discussed in Section 5.1). In our work, we tackled this risk by making the interviews as informal as possible through appropriate ways of communication and interaction with the participants.

RELATED WORK

This section describes related work on software ecosystems and initially provides descriptions of previous attempts in defining the

dependent on hardware, and contain other characteristics, such as safety requirements, that lead to further challenges for an ecosystem.

CONCLUSIONS

In this paper, we have described a case study of the ecosystems for FES. The basis for the analysis was a series of interviews with representatives of the key actors in such an ecosystem. In all our interviews the discussion led to a clear indication that the benefit from letting ES connect and allowing dynamic software add-ons is very interesting for the industries included in the case study, especially when it comes to providing new functionality and services, reducing time to market, adapting to niche segments, and improving innovation through working with external partners. However, the interviews also clearly confirmed our hypothesis that it is not sufficient to only solve the technical challenges with FES, but that business related issues are at least as important and demanding. A more comprehensive view of our findings relates to our hypothesis and is thus considered of special interest.

Summary of findings

In the paper, two RQs were posed and we will now briefly summarize the answers to them.

1. *What actors are needed in an ecosystem of Federated Embedded Systems (FES), and what are their relations?*

As it was presented in Section 3.1, a large number of actors are present in the ecosystems of FES. The actors cover the traditional supply chain of ES, including suppliers, manufacturers, owners, and end users. However, many new actors are added, such as content providers, communication providers, service operators, add-on developers, and information brokers. The roles and responsibilities of each actor also changes. As a striking example, the manufacturer is no longer responsible for the final integration of the product. Instead, the owner or user adds additional software to the product, from third-party add-on developers that might not even have formal relations to the manufacturer. This leads to many challenges related to development methods, liability, etc.

2. *What are the key elements of (a) the business models, (b) the product architecture, and (c) the processes, methods, and tools, needed to make the ecosystem effective?*

The business model of a FES ecosystem was analyzed using the BMC, and the analysis showed many new ways of doing business compared to what is traditional for ES, including both new value propositions from manufacturers and other actors, as well as adapting to new customer segments. The revenue streams for non-traditional ES particularly include many new ways of selling software products and information. All in all, the application of FES could in many ways reshape the ES industry.

It is clear from the study that the architecture is a key success factor for FES, and it is not limited to the ES architecture of the product, but to the overall enterprise architectures of the actors in the ecosystem. The architecture also reflects some of the important trade-offs that are a consequence of the FES concept, such as openness vs. dependability, which is due to the safety-critical nature of many ES, and the need for hardware cost efficiency. All these are factors that differentiate the FES ecosystem from pure software ecosystems.

Finally, the processes, methods, and tools for developing ES need to be revised as well, and in particular the interplay between manufacturers and add-on developers must be defined. This includes both what support an add-on developer can expect from the manufacturer, and what the responsibilities are for ensuring quality of the end product. To some extent, manufacturers will try to retain control, but this will again be a trade-off against the openness of the ecosystem.

Future work

From the data collected in this study, many important challenges can be identified that provide topics for further research, including the following:

- Development of the technical mechanisms for handling plug-ins, in accordance with the requirements identified in this study, such as dependability, cost, etc.
- Conduct further case studies based on the conceptual model generated in this research, to assert to what extent it can be generalized to other industry segments and to understand important limitations.
- Characterization of an appropriate software architecture, e.g., how to structure the API, how to deal with conflicts between plug-ins, and how to efficiently create software systems on the federation level.
- Experimental verification of the open innovation aspects, by development of add-on software in a controlled or semi-controlled environment.
- Refinement of the BMC by creating an individual BMC for each actor type, and make the links between them more explicit.

ACKNOWLEDGEMENTS

The authors would like to express their thanks to all participants in this study. The research was funded by VINNOVA, the Swedish Agency for Innovation Systems (Grant No. 2012-03782). The work of Dr. Papatheocharous was carried out during the tenure of an ERCIM "Alain Bensoussan" Fellowship Programme. The research leading to these results has received funding from the European Union Seventh Framework Programme(FP7/2007-2013) under Grant Agreement No. 246016.

APPENDIX A. INTERVIEW CHECKLIST

This appendix presents the checklist used during the interviews. Note that it was not used as a questionnaire, and thus it does not reflect the flow of the interviews conducted. Rather, interviewers used it as a way of ensuring that all important topics were covered at some degree and at some point during the interviews. The interviews were mainly carried out as open dialogues.

A. Preliminary data

1. Briefly describe the background of this research and purpose of the interview.
2. Discuss the interviewee's role at the company.
3. What are the total number of years working in the industry and number of years in the company?

B.Roles and relations in ecosystem

4. What type of external organizations do you interact with (i.e., Original Equipment Manufacturers (OEM), suppliers, add-on developers, service providers, others)?
5. What is your motivation for working with these partners?
6. Who typically initiates the interactions around an idea of a new product? You? A partner?
7. Have you experienced any conflicts as a consequence of working with external partners?
8. How do you structure your products to work efficiently with partners? Product lines?

C.Business models

9. What value do you offer to your customers, e.g. what need is satisfied, or problem solved?
10. Do you consider satisfying any particular customer segments, or types of customers?

11. In what form do you get revenue from the customers? Per installation, per use, for free, etc.?
12. What are the business models used in the interaction with these organizations, i.e., how do you share revenues and costs with them?
13. How are products and add-ons sold and distributed to the customers (end-users)? Are joint or separate channels used?
14. How do you maintain a relation over time with customers, e.g., for making updates, providing associated services, etc.?

D.Development processes, methods and tools

15. What type of information do you give to, or receive from, these organizations during development?
16. How does working with external partners influence the way you structure your technology (product architecture, interfaces, etc.)?
17. Who is responsible for the quality of the total solution, and how are liability issues resolved?

E.Wrap-up

Is there anything else that we should have discussed?

REFERENCES

1. U. Eklund, J. Bosch, Introducing software ecosystems for mass-produced embedded systems, Softw. Business (2012) 248–254.
2. J. Axelsson, A. Kobetski, On the conceptual design of a dynamic component model for reconfigurable AUTOSAR systems, in: Proceedings of the 5th Workshop on Adaptive and Reconfigurable Embedded Systems (APRES'13), Philadelphia, USA, April 8-13, 2013.
3. D.G. Messerschmidt, C. Szyperski, Software Ecosystems: Understanding an Indispensable Technology and Industry, MIT Press, 2005.

4. R.P. Dos Santos, C.M.L. Werner, Revisiting the concept of components in software engineering from a software ecosystem perspective, in: Proceedings of the Fourth European Conference on Software Architecture: Companion Volume, 135–142, ECSA'10, ACM, New York, NY, USA, 2010. doi:http:// dx.doi.org/10.1145/ 1842752. 1842782.
5. R.K. Yin, Case Study Research: Design and Methods, fifth ed., SAGE Publications, 2013.
6. C. Robson, Real World Research, second ed., Blackwell, 2002.
7. P. Runeson, M. Höst, Guidelines for conducting and reporting case study research in software engineering, Empir. Softw. Eng. 14 (2) (2009) 131–164.
8. A. Osterwalder, Y. Pigneur, Business Model Generation, John Wiley, 2010.
9. K. Manikas, K.M. Hansen, Software ecosystems – a systematic literature review, J. Syst. Softw. 86 (5) (2013) 1294–1306, http://dx.doi.org/10.1016/ j.jss.2012.12.026.
10. European Union, Directive 2006/42/EC, Official Journal of the European Union, pp. L157/24-86, 2006.
11. G.K. Hanssen, T. Dybå, Theoretical foundations of software ecosystems, in: Proceedings of the 4th Third International Workshop on Software Ecosystems (IWSECO'11), Boston, Massachusetts, 2011, pp. 6–17.
12. R. Goldman, R.P. Gabriel, Innovation happens elsewhere open source as business strategy, Morgan Kaufmann, Amsterdam; Boston, 2005.
13. C. Wohlin, P. Runeson, M. Höst, M. Ohlsson, B. Regnell, A. Wesslén, Experimentation in Software Engineering: An Introduction, Springer, 2000.
14. K. Popp, Goals of software vendors for partner ecosystems – a practitioner's view, Softw. Business (2010) 181–186.
15. E. Yu, S. Deng, Understanding software ecosystems: a strategic modeling approach, in: Proceedings of the 3rd International Workshop on Software Ecosystems (IWSECO'11), Brussels, Belgium, CEUR: 65–76, 2010. .
16. E. Papatheocharous, J. Axelsson, J. Andrersson, Issues and challenges in ecosystems for federated embedded systems, in: Proceedings of the First International Workshop on Software Engineering for Systems-of-Systems (SESoS'13), Montpellier, Paris, 21–24, ACM, New York, NY, USA, 2013. doi:http://dx.doi.org/10.1145/2489850.2489854.
17. J. Bosch, From software product lines to software ecosystems, in: Proceedings of 13th International Software Product Line Conference (SPLC'09), IEEE Computer Society, San Francisco, USA, 2009, pp. 111–119.

18. J. Bosch, P. Bosch-Sijtsema, From integration to composition. On the impact of software product lines global development and ecosystems, J. Syst. Softw. 83 (2010) 67–76.
19. V. Boucharas, S. Jansen, S. Brinkkemper, Formalizing software ecosystem modeling, in: Proceedings of 1st International Conference on Open Component Ecosystems (IWOCE'09), ACM, Amsterdam, 2009, pp. 41–50.
20. A. Ghazawneh, O. Henfridsson, Micro-strategizing in platform ecosystems: a multiple case study, in: Proceedings of the 32nd International Conference on Information Systems (ICIS 2011), Shanghai, China, December 4–7, 2011.
21. S. Jansen, A. Finkelstein, S. Brinkkemper, A sense of community: a research agenda for software ecosystems, in: Proceedings of the 31st International Conference on Software Engineering, New and Emerging Research Track, 2009, pp. 187–190.
22. O. Barbosa, C. Alves, A systematic mapping study on software ecosystems, in: Proceedings of the 3rd International Workshop on Software Ecosystems (IWSECO'11), Brussels, Belgium, 2011. .
23. S. Jansen, M.A. Cusumano, S. Brinkkemper, Software Ecosystems: Analyzing and Managing Business Networks in the Software Industry, Edward Elgar Publishing, 2013.
24. G.K. Hanssen, A longitudinal case study of an emerging software ecosystem: implications for practice and theory, J. Syst. Softw. 85 (7) (2012) 1455–1466.
25. M. Iansiti, R. Levien, Strategy as ecology, Harvard Bus. Rev. 82 (3) (2004) 68- 81.
26. J. Bosch, P.B. Sijtsema, From integration to composition: on the impact of software product lines, global development and ecosystems, J. Syst. Softw. (2010) 67–76.
27. T. Kilamo, I. Hammouda, T. Mikkonen, T. Aaltonen, From proprietary to open source—growing an open source ecosystem, J. Syst. Softw. 85 (7) (2012) 1467- 1478.
28. I. Hunink, R. Van Erk, S. Jansen, S. Brinkkemper, Industry taxonomy engineering: the case of the European software ecosystem, in: Proceedings of the 4th European Conference on Software Architecture (ECSA 2010), ACM, Copenhagen, Denmark, August 23–26, 2010, pp. 111–118.
29. I.V.D. van de Berk, S. Jansen, L. Luinenburg, Software ecosystems: a software ecosystem strategy assessment model, in: 2nd International Workshop on Software Ecosystems (EcoSys'10), Proceedings of the 4th European Conference on Software Architecture (ECSA'10), Companion Volume, Copenhagen, Denmark, August 23–26, 2010, pp. 127–134.

30. S. Jansen, S. Brinkkemper, A. Finkelstein, Business network management as a survival strategy: a tale of two software ecosystems, in: Proceedings of the First Workshop on Software Ecosystems, 2009, pp. 34–48.
31. S. Jansen, S. Brinkkemper, A. Finkelstein, Providing transparency in the business of software: a modeling technique for software supply networks, in: 8th IFIP Working Conference on Virtual Enterprises (Pro-VE'07), Guimarães, Portugal, September 10–12, 2007.
32. B. Iyer, C.-H. Lee, N. Venkatraman, Managing in a "small world ecosystem": lessons from the software sector, Calif. Manage. Rev. 48 (3) (2006) 28–47.

CITATION

Jakob Axelsson, Efi Papatheocharous, Jesper Andersson, Characteristics of software ecosystems for Federated Embedded Systems: A case study, Information and Software Technology, Volume 56, Issue 11, November 2014, Pages 1457-1475, ISSN 0950-5849, http://dx.doi.org/10.1016/j.infsof.2014.03.011.

CHAPTER 5

Embedded Strain Gauges for Condition Monitoring of Silicone Gaskets

*Timo Schotzko and Walter Lang **

Institute for Microsensors, Actuators and Systems (IMSAS), University of Bremen, Otto-Hahn-Allee NW1, 28359 Bremen, Germany; E-Mail: tschotzko@imsas.uni-bremen.de

ABSTRACT

A miniaturized strain gauge with a thickness of 5 μm is molded into a silicone O-ring. This is a first step toward embedding sensors in gaskets for structural health monitoring. The signal of the integrated sensor exhibits a linear correlation with the contact pressure of the O-ring. This affords the opportunity to monitor the gasket condition during installation. Thus, damages caused by faulty assembly can be detected instantly, and early failures, with their associated consequences, can be prevented. Through the embedded strain gauge, the contact pressure applied to the gasket can be directly measured. Excessive pressure and incorrect positioning of the gasket can cause structural damage to the material of the gasket, which can lead to an early outage. A platinum strain gauge is fabricated on a thin polyimide layer and is contacted through gold connections. The measured resistance pressure response exhibits hysteresis for the first few strain cycles, followed by a linear behavior. The short-term impact of the embedded sensor on the stability of the gasket is investigated. Pull-tests with O-rings and test specimens have indicated that the integration of the miniaturized sensors has no negative impact on the stability in the short term.

INTRODUCTION

Rubber and silicone gaskets are usually small yet crucial components in many technical systems. A gasket is usually a low-cost product and comprises only a small part of the overall cost of the total system. However, its role in safety and long-term function is significant. Defective seals can sometimes result in disproportionate failures; a classic example is the explosion of the Space Shuttle Challenger in 1986, which was caused by the failure of a rubber O-ring [1]. At present, it is impossible to directly monitor the condition of the gasket in its in-service state; hence, reaction to a possible failure or leakage cannot be performed preemptively. Consequently, gaskets must be periodically replaced to avoid failures, especially in security-critical applications. This schedule-based rather than needs-based approach leads to extra effort and expense. One possibility for monitoring the state of an O-ring is to embed sensors into the gasket.

Various factors can affect the functionality of rubber gaskets. In addition to ordinary aging of the materials, degradation can be hastened by contact with incompatible chemicals and gases and by unsuitable temperatures. The elastomeric materials become porous and a transition from elastic to viscoelastic behavior occurs. As a consequence, the compression set increases, and the gasket retains its shape even after pressure has been released. The reset force of the gasket decreases and leakage occurs. Another reason for a failure may be from an incorrect assembly or fitting. A faulty gasket and uneven or excessive compression often lead to malfunctioning. On the other hand, insufficient compression can directly lead to leakages [2].

O-rings are found in many applications. Depending on usage, a defective gasket can present a high safety risk or danger to the environment, e.g., in areas where hazardous chemicals and gases are sealed. A malfunction of these low-cost components can also produce disproportionately high costs. Many rubber gaskets, for example, are built into pneumatic and hydraulic systems, and leakages affect the energy efficiency of such systems. A study from 2001 about "compressed air systems in the European Union" has shown that high energy losses occur due to lack of preventive maintenance. The study assumes that prevention of air leaks could save up to 12.8 billion kWh a year in the European Union (in 2001). This corresponds to 16% of the total energy consumption in this field [3]. Therefore, monitoring the condition of single gaskets is of great interest because the occurrence of leakages could be diminished and the need for extensive preventive maintenance could be avoided.

As an example for a self-monitoring gasket, this study presents the design and performance of an embedded strain gauge for O-ring monitoring. Through this technology, the deformation of a gasket can be measured, and the contact pressure determined. The integration of a strain gauge could provide a promising basis to embed additional sensors in silicone and rubber materials, including temperature sensors and sensors to measure dielectric permittivity.

Structural Health Monitoring

Sensor integration has become an important research topic over the past decade. Structural health monitoring (SHM) is a highly discussed topic and continues to attract interest in different fields. The areas of application have been enlarged in recent years, and systems have become increasingly complex. Since the late 1980s, this idea has been established in the fields of aerospace, civil, and mechanical engineering [4–6]. The aim of SHM is to provide a continuous diagnosis of the state of the controlled material and structure in a nondestructive manner. Thus, defective materials, damage from external influences, and aging processes can be monitored and evaluated. The results can be used to provide a diagnosis of the development of damage and a prognosis of the remaining service life [7]. Embedding sensors in O-rings, as a part of SHM systems, is a new approach of great interest to many industrial applications.

The integration of sensors presents two major challenges. On the one hand, the sensors should be designed such that the host material will not be damaged, although any embedded structure always represents an interruption to the material. On the other hand, it is also important that the sensor provides a reliable, repeatable, and preferably linear response to the parameters being measured.

Therefore, the sensors must be designed with a size and choice of material such that it does not weaken the structure of the gasket, but still provides the information needed to monitor the condition [8]. One possibility to minimize the influence of the embedded sensor is to select a material for the sensor substrate, which is chemically similar to the host material. Thus, it is possible to enable cross linking between sensor and gasket, as shown in Figure 1.

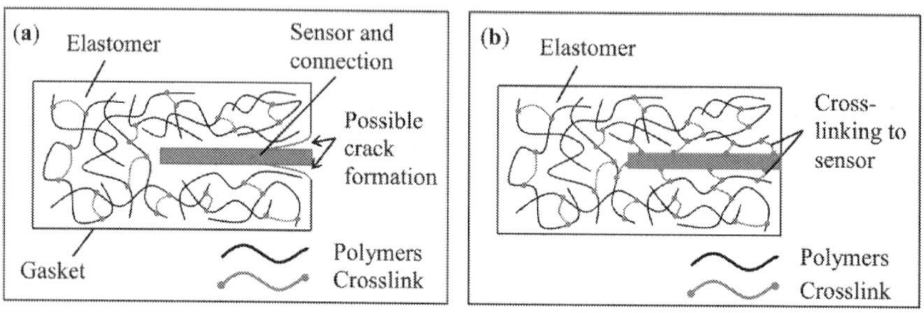

Figure 1. A sketch of an embedded sensor in a gasket (**a**) with no crosslinking agents between sensor and elastomer, which exhibits a possible source of crack formation and separation; (**b**) with crosslinking between host material and sensor structures.

State of the Art

Sensor Integration

Sensor integration together with SHM is a generally topical subject. The data collected by integrated sensors can provide various types of information about the host material. The evaluation of the received data is of great interest for research and development. It can reveal the behavior and changing properties of a host material and a hosting device online. This allows us to gain greater comprehension of materials themselves and of the whole system. It also enables several new possibilities of analysis for monitoring during implementation as well as testing the properties of materials and devices. For gaskets, it could also complement standard rubber testing methods with applicable sensors. For example, it would be possible to measure the chemical change in the rubber material during exposure to harmful substances. Through the embedded nature of the sensors, a more precise evaluation of the impact characteristics of the material would be possible.

In recent years, different sensors have been successfully integrated in various materials, such as metals [9–11], textiles [12,13], and fiber-reinforced polymers (FRPs) [5]. Temperature sensors have been embedded in aluminum alloys to detect temperatures during molding processes [9]. Other groups have integrated thermoelectric energy generators [10] and a radio-frequency identification transponder [11] in aluminum by casting. The integration of sensors in metals is highly challenging because of the harsh conditions during metal casting, rolling, and forming processes.

In contrast to the challenges associated with embedding sensors in structural materials are those associated with embedding sensors into textiles. These materials are usually very flexible and stretchable. Therefore, the sensors have to be compliant and operate over wide strains. They must also be resistant to chemicals, soaps, and perspiration. The embedding processes are different in this case as well. Instead of molding, the sensors must be woven in the materials. Common usage of integrated sensors and electronics is in smart textiles for sports, lifestyle, and the health industry. Clothing that includes sensors for monitoring an electrocardiogram [12] and shoes with energy-harvesting microelectronics have already been demonstrated [13], among many other examples.

FRPs and carbon-fiber-reinforced polymers (CFRPs) are commonly used as rugged and lightweight materials in aircrafts and in wind turbine blades. Possible applications for integrated sensors are measuring acting forces, monitoring fiber breakage, and observing the change in permittivity at the interface between the sensor and the host material, caused by polymerization within the manufacturing process. To avoid delamination, the sensors have to be miniaturized as well as flexible. Integration into the CFRP of a flexible sensor based on polyimide, with a thickness of about 5 μm, has already been demonstrated [14].

Common Gasket Leakage Detection Methods

Leakages are commonly detected by indirect methods. Different pressure and flow sensors are usually integrated in industrial facilities to indirectly monitor the symptoms of leakage rather than directly monitor early gasket failure. Thus, a loss in process pressure and vacuum can be detected, and a leakage rate can be calculated. A significant disadvantage of this method is that the exact location of the defective gasket is not directly indicated [15,16]. Therefore, these types of monitoring approaches are rarely implemented in industrial facilities. The most prevalent method for directly finding a defective gasket is performed manually using an ultrasound detector. The detector can identify the ultrasound emission of a leak, thus enabling localization of the leakage at the site of the faulty gasket [17].

A different approach for detecting faulty gaskets is through external gas sensors, usually applied in chemical process plants. In an emergency, if poisonous or inflammable gas leaks, the plant can be shut down to find the leakage. In addition, other methods exist to detect leakage of liquids through defective gaskets. Seals with an interior repository can collect leaking liquids, such as oil, that are detectable [18]. However, all the above mentioned leakage-detection methods require significant installation effort, are bulky, and are associated with high expense.

Gaskets with Sensors

Gaskets with sensors have been investigated for considerable time and via many different methods, some of which have proved successful. One of the most widely known is the radial shaft seal with an integrated leakage sensor, which is currently used in wind turbines [18]. These gaskets detect leaked liquids (normally oils) and are usually made of an interconnected system of metal and rubber. The condition of the gasket is indirectly monitored by detecting oil in a depot inside the gasket using an optical sensor. The range of applications is limited for this type of gasket. Other examples are cylinder head gaskets with integrated temperature sensors, which enable optimized engine management [19] as well as a plastic gasket with a sensor array on it. The sensors detect fault assembling, leakages, and chemical and physical aging processes of the gaskets [20].

Concept of Condition Monitoring for Gaskets by Using Strain Gauges

The motivation of the current study is to obtain as much information as possible about the condition of a gasket through embedded sensors. Various parameters, such as compression set, permittivity, and contact pressure, could provide relevant information. In this article, a strain gauge is described as a gauge that measures the contact pressure of the gasket. The concept is to integrate a strain gauge centered within an O-ring, so that radial strain can be detected by the sensor. The strain arises from vertical pressure applied to the gasket. The embedded strain gauge is elongated under the gasket contact pressure, which leads to a measurable increase in resistance. The connection is made with gold conducting paths on the same thin and flexible substrate as the sensor. These lead outside the gasket, so that they can be connected to a measurement unit. Thereby, the resistances of the contact paths are much lower than that of the strain gauge structure to minimize loading of the measurement signal. The structure of the embedded sensor in the gasket and the connections are shown in Figure 2.

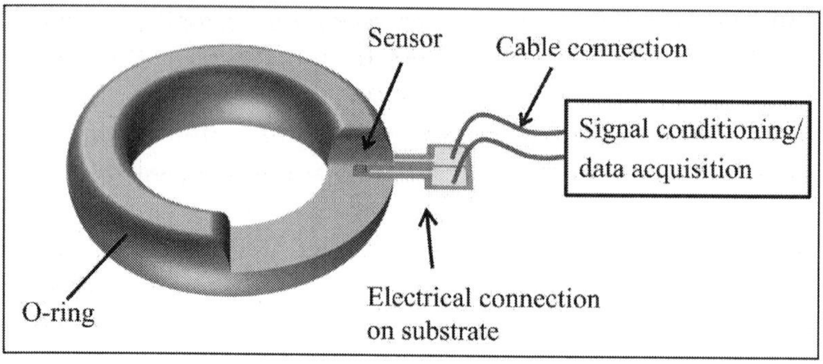

Figure 2. Illustration of an embedded strain gauge in a silicone gasket.

EXPERIMENTAL SECTION

The strain gauges are processed as a thin-film platinum resistor on polyimide. A variety of different meander structures are implemented. Gold is used for interconnections on the polyimide foils. Subsequently, the sensors are molded into silicone O-rings and examined. An overview of the single- processing steps of the sensors is presented in Figure 3.

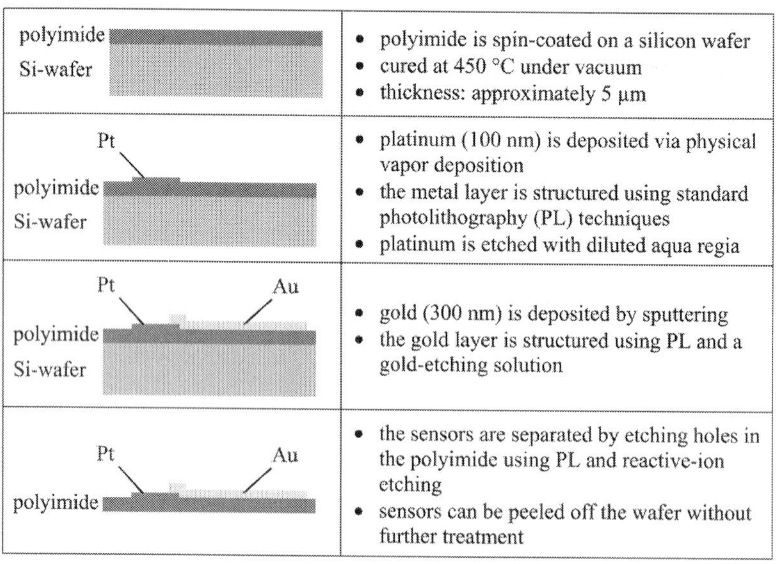

Figure 3. Summary of the processing steps of the sensor.

A 5-µm-thick polyimide layer is spin-coated on a silicon wafer and annealed under vacuum. Platinum is deposited on the polyimide and structured by standard photolithography (PL) techniques and diluted aqua regia. In addition, the gold conductive paths and pads are sputtered on and patterned by PL. The single strain gauges are separated by etching the polyimide foil using reactive-ion etching. Because of the annealing process of the polyimide on silicon, the sensors (see Figure 4) can be simply peeled off from the silicon wafer with no further treatment.

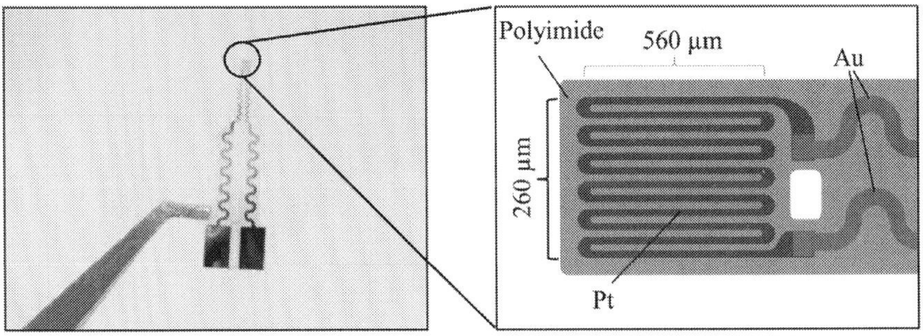

Figure 4. Photograph of a strain gauge and schematic of the meander structure.

In the second step, the gaskets and test specimens are fabricated. For this, a casting mold with a bracket for the sensors is used to embed the strain gauges. The gauge pattern is located inside the casting mold when liquid silicone is molded. One strain gauge is embedded in each O-ring and each test specimen, as shown in Figure 5. The silicone gasket is cured for 20 min at 120 °C.

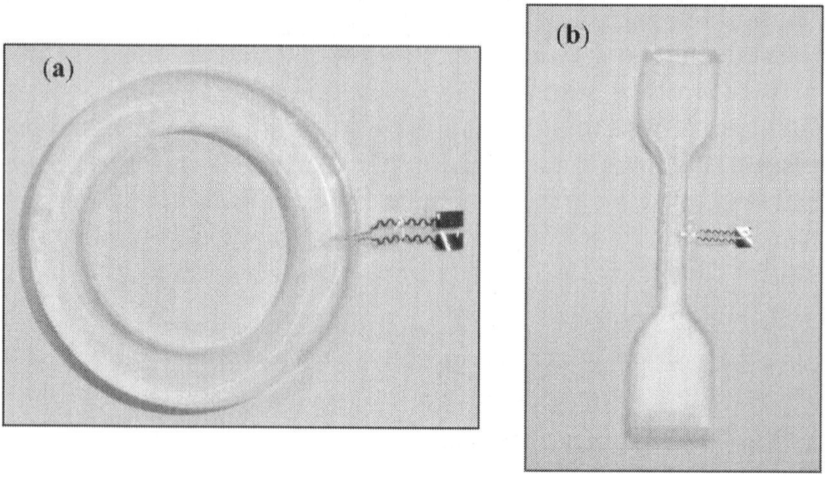

Figure 5. Photographs of an embedded sensor in (**a**) an O-ring and (**b**) a test specimen.

Sensor Validation and Performance Characterization

Validation of Embedded Sensor

The prepared gaskets are placed between two metal plates in a force-measurement unit (XYZTEC Condor with pull-and-push measurement unit) as shown in Figure 6a,c. The sensors are connected by two cables to a resistance meter (HM8012 programmable multimeter) while the gauge head applies pressure on the O-rings. The measurement head is moved in 100 μm steps in a range of 0–1500 μm. The force of the applied pressure is monitored and correlated with the resistance of the strain gauge.

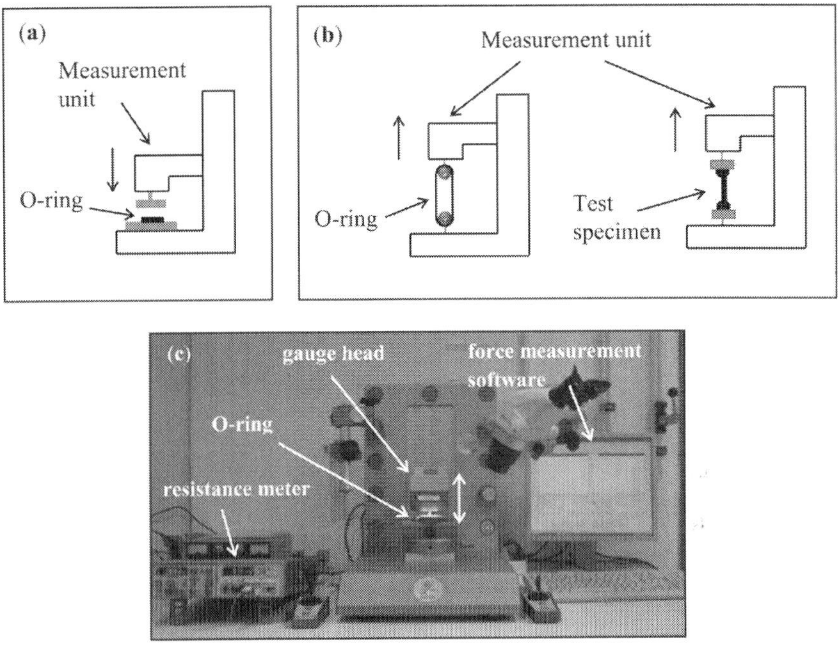

Figure 6. Sketches of the measurement setup. (**a**) for pressure tests; (**b**) pull-tests; (**c**) photograph of the measurement setup for pressure test.

Uniaxial Tensile Testing on Dog-Bone Test Specimens

With the same measurement unit, we also perform tensile testing (Figure 6b). Therefore, dog-bone test specimens are produced with and without embedded sensors. The strain gauges are centrally placed within the test specimens, as illustrated in Figure 5b. Subsequently, they are tested in uniaxial tensile to failure. The elongation of the specimens is recorded against the applied force. The fracture strengths of the dog-bone test specimens with and without embedded sensors are compared.

Uniaxial Tensile Testing on O-rings

Uniaxial tensile testing is also performed on the O-rings using suitable hooks (Figure 6b). In this case, the fracture patterns of the O-rings with integrated sensors are evaluated to determine whether the sensor affects the point of fracture compared with O-rings without sensors. On the basis of the tensile testing, a statement can be made about a possible influence of the embedded sensors on the performance stability of the gasket.

RESULTS AND DISCUSSION

The produced strain gauges have resistances between 600 and 3300 Ω because of the different length, width, height, and distance of the strain gauge patterns. Currently, no evidence exists that the curing process during the embedding into silicone has any influence on the sensor resistance.

Force-Resistance Measurements

The force-resistance response of the embedded sensor exhibits linear behavior at a constant temperature (room temperature), as illustrated in Figure 7a.

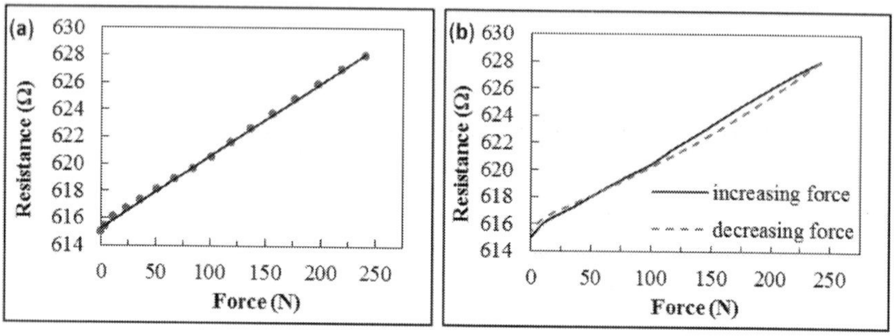

Figure 7. Resistance-force characteristic of a strain gauge embedded in a silicone O-ring. (**a**) Nearly linear response while increasing force; (**b**) Hysteresis during the first strain cycle.

At low forces (<20 N), the graph shows a slight deviation, caused by the surface characteristics of the O-rings. As a result of the production process, the gasket is slightly thicker around the embedded sensor. Hence, the O-ring surface touches the measurement unit unevenly. The contact pressure is not applied uniformly across the O-ring at the beginning of the measurement. This leads to an irregular strain distribution that causes the observed deviation in resistance of a strictly linear response. However, as soon as the metal plate is consistently pressed onto the gasket (usually at about 20 N contact force), the measured data exhibits a proportional relationship between contact force and resistance.

Although the strain gauge exhibits linear behavior in response to increased applied force, a hysteresis effect was observed during the removal of force from the gasket (see Figure 7b). At high contact forces, the measured resistance of the strain gauge is lower when force is being reduced. However, with further reduction in force, the decrease lessens, with the result that a zero point offset of the resistance occurs toward higher values after unloading the strain gauge. This performance can be observed during the first four cycles. The hysteresis effect decreases each time, until there is almost no difference in behavior during cycling of the force, as shown in Figure 8. The zero offset is also negligible.

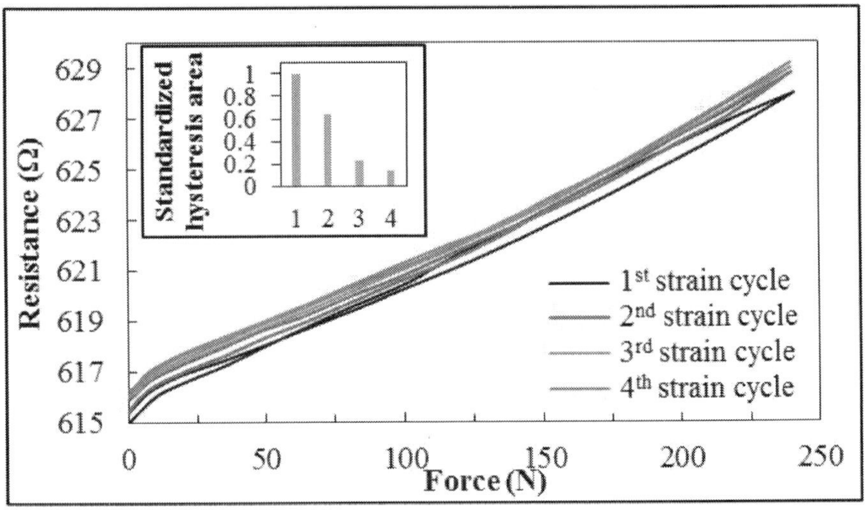

Figure 8. Hysteresis effect of resistance-force response during first four cycles of loading and unloading the strain gauge. The bar chart points out the change in area of the hysteresis loops.

Hysteresis effects are already well described for strain gauges. These effects can be caused by cold hardening of the metals, properties of adhesion to the substrate material, or, in this case, the host material of the strain gauge foil causes hysteresis as well as the zero offset [21–23]. The hysteresis depends on the maximum applied strain and the number of cycles [21], as observed during these measurements. Another possible reason for the hysteresis is that during the curing process of the silicone, residual tensile strains are locked in the O-ring. Alternating strain and relaxation could relieve these residual strains and lead to a minor creep deformation of the silicone followed by a zero offset.

Uniaxial Tensile Testing

Preliminary results have shown that the integration of small sensors does not influence the stability of silicone O-rings. However, long-term measurements have not yet been performed. Accordingly, a negative influence over time through introduction of the sensors cannot be ruled out at this point.

Investigations to indicate whether the embedded strain gauges weaken the structural stability of the O-rings showed that no negative impact caused by the sensors could be observed. Tensile testing is performed on the gaskets and the test specimens, both with and without embedded sensors. The fracture strength is detected by the same measurement unit mentioned in Section 2.1.

Test Specimens

The results of the tensile testing show no differences between the dog-bone specimens with or without an embedded sensor. A range of measurement data is shown in Figure 9. The average fracture strength of all measurements is 1.44 N/mm^2 (σ = 0.38 N/mm^2) without sensors and 1.48 N/mm^2 (σ = 0.29 N/mm^2) with the embedded sensor.

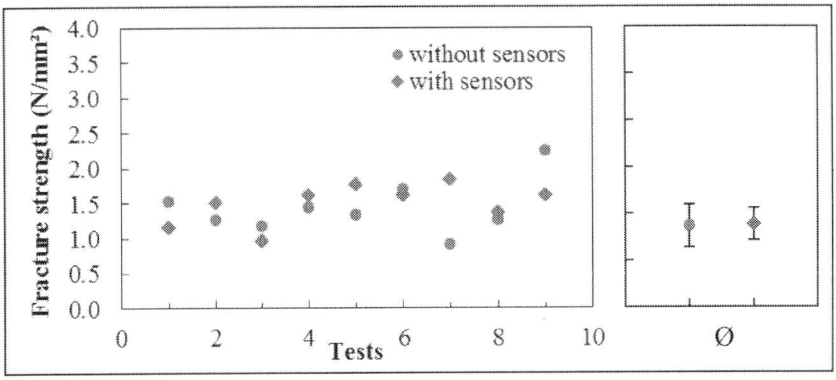

Figure 9. Comparison of the tensile testing results of the dog-bone test specimens with and without integrated sensor.

O-Rings

Uniaxial tensile testing with O-rings gives the same results. No differences are observed between silicone O-rings with or without an embedded sensor, neither by destructive forces nor by rupture location. The gaskets tear apart randomly and not specifically where the strain gauge is located. This shows that the sensors do not influence the integrity of silicone gaskets in the short term.

CONCLUSIONS AND OUTLOOK

A first approach in integrating strain gauges in silicone gaskets has been shown. The relationship between the contact force and the sensor resistance has been analyzed. A nearly linear behavior between the output signal and the force was observed, although hysteresis was observed during the first few cycles of contact-force application. Tensile testing has indicated that there are no negative effects on the stability of the gasket caused by the embedded sensors, at least in the short term.

These results are very promising for further research on self-monitoring gaskets. They provide the basis for integrating various sensors for condition monitoring of gaskets. Future work should be implemented to examine the long-term effects of embedded sensors on the host materials. Further, the sensor itself can be improved using different substrate materials to support an enhanced bond between the integrated structures and gasket material. In this context, other materials that can be linked through covalent bonds to the host material can be investigated to explore positive impacts on the stability of the gasket.

In future work, we will investigate methods to detect potential degradation of gaskets. Degradation changes the elastic behavior from elastic to viscoelastic, which in turn causes a loss of sealing pressure. It is expected that the change in elastic behavior is accompanied by deformations that are small but measurable by strain gauges. However, the experimental validation is yet to be performed and is a topic of future research work.

AUTHOR CONTRIBUTIONS

The work presented in this paper was a collaboration of both Authors. Walter Lang and Timo Schotzko developed the idea and design of the integrated sensors. Timo Schotzko processed and integrated the strain gauges. He also performed the measurements and characterization as well as writing the manuscript, which was discussed and revised by both authors.

REFERENCES

1. Presidential Commission. Report of the Presidential Commission on the Space Shuttle Challenger Accident; Report to the President. NASA History Office: Washington, DC, USA; June; 1986.
2. Berk, J. Leaks. Systems Failure Analysis; ASM International: Materials Park, OH, USA, 2009; pp. 131–139.
3. Radgen, P.; Blaustein, E. Compressed Air Systems in the European Union Energy, Emissions, Savings Potential and Policy Actions; LOG_X Verlag GmbH: Stuttgart, Germany, 2001.
4. Antunes, P.C.; Dias, J.M.; Varum, H.; André, P. Dynamic Structural Health Monitoring of a Civil Engineering Structure with a Pof Accelerometer. Sens. Rev.**2014**, 34, 36–41.

5. Kahandawa, G.C.; Epaarachchi, J.; Wang, H.; Lau, K.T. Use of FBG Sensors for SHM in Aerospace Structures. Photonic Sens. **2012**, 2, 203–214.
6. Adams, D.; White, J.; Rumsey, M.; Farrar, C. Structural Health Monitoring of Wind Turbines: Method and Application to a Hawt. Wind Ener. **2011**, 14, 603–623.
7. Güemes, A.; Fritzen, C.-P.; Balageas, D. Introduction to Structural Health Monitoring. In Structural Health Monitoring; ISTE: London, UK, 2006; pp. 13–44.
8. Dumstorff, G.; Paul, S.; Lang, W. Integration without Disruption: The Basic Challenge of Sensor Integration. IEEE Sens. J. **2014**. [CrossRef]
9. Cheng, X.D.; Datta, A.; Choi, H.; Zhang, X.G.; Li, X.C. Study on Embedding and Integration of Microsensors into Metal Structures for Manufacturing Applications. J. Manuf. Sci. Eng. **2007**, 129, 416–424.
10. Ibragimov, A.; Pleteit, H.; Pille, C.; Lang, W. A ThermoelectricEnergy Harvester Directly Embedded into Casted Aluminum. IEEE Electr. Dev. Lett. **2012**, 33, 233–235.
11. Pille, C. In-Process Embedding of Piezo Sensors and Rfid Transponders into Cast Parts for Autonomous Manufacturing Logistics. In Smart Systems Integration, Proceedings of 4th European Conference & Exhibition on Integration Issues of Miniaturized Systems—MEMS, MOEMS, ICs and Electronic Components, Como, Italy, 23–24 March 2010.
12. Catrysse, M.; Puers, R.; Hertleer, C.; Van Langenhove, L.; van Egmond, H.; Matthys, D. Towards the Integration of Textile Sensors in a Wireless Monitoring Suit. Sens. Actuator A Phys. **2004**, 114, 302–311.
13. Kymissis, J.; Kendall, C.; Paradiso, J.; Gershenfeld, N. Parasitic Power Harvesting in Shoes. Proceedings of Second International Symposium on Wearable Computers, Pittsburgh, PA, USA, 19 October 1998.
14. Boll, D.; Tolstosheeva, E.; Schubert, K.; Brauner, C.; Lang, W. Flexible Interdigital Sensor for Dielectric Cure Monitoring of Composites. Proceedings of DEWEK 2012 German Wind Energy Conference, Bremen, Germany, 7–8 November 2012.
15. Fraunhofer ISI; dena; VDMA. Compressed Air Facts. Druckluft Effizient **2003**.
16. Nakutis, Z.; Kaskonas, P. An Approach to Pneumatic Cylinder on-Line Conditions Monitoring. Mechanika **2008**, 4, 41–47.
17. SDT International s.a. Guide for Ultrasonic Leak Detection in Compressed Air Systems in an Industrial Environment, 2nd ed.; SDT International s.a.: Brussels, Belgium, 2014.

18. Freudenberg Simrit GmbH & Co. KG. Simmerrring Mss1+ Condition Monitoring—Safety and Reliability through Innovative Seal Function Monitoring. Available online: http://www.eriks.nl/documentatie/dynamische-afdichtingen/simrit-documentatie/simmerring-oil-seals-condition-monitoring.pdf (accessed on 9 July 2014).
19. Laske, M.; Ludwig, J.; Unseld, G.; Weiß, A. Cylinder Head Gaskets with Integrated Temperature Sensors. MTZ Worldw. **2002**, 63, 2–5.
20. Kremer, F. Method for Determining Al Least One State Parameter of a Sealing System and Sealing System. Patent EP1597554B1, 2008.
21. Sciammarella, C.A.; Sciammarella, F.M. Experimental Mechanics of Solids; Wiley: Chichester, UK, 2012.
22. Hannah, R.L.; Reed, S. Strain Gage Users' Handbook, 1st ed.; Elsevier Science Publishers Ltd.: Cambridge, UK, 1992.
23. Klaassen, K.B. Electronic Measurement and Instrumentation; Cambridge University Press: Cambridge, UK, 1996.

CITATION

Timo Schotzko and Walter Lang; Embedded Strain Gauges for Condition Monitoring of Silicone Gaskets; doi:10.3390/s140712387.

CHAPTER 6

A Vision-Based Driver Nighttime Assistance and Surveillance System Based on Intelligent Image Sensing Techniques and a Heterogamous Dual-Core Embedded System Architecture

Yen-Lin Chen[1], *Hsin-Han Chiang*[2], *Chuan-Yen Chiang*[3], *Chuan-Ming Liu*[1], *Shyan-Ming Yuan*[3] *and Jenq-Haur Wang*[1,*]

[1]Department of Computer Science and Information Engineering, National Taipei University of Technology, 1, Sec. 3, Chung-hsiao E. Rd., Taipei 10608, Taiwan;
[2]Department of Electrical Engineering, Fu Jen Catholic University, New Taipei City 24205, Taiwan;
[3]Department of Computer Science, National Chiao Tung University, 1001 University Road, Hsinchu 30050, Taiwan;

ABSTRACT

This study proposes a vision-based intelligentnighttime driver assistance and surveillance system (VIDASS system) implemented by a set of embedded software components and modules, and integrates these modules to accomplish a component-based system framework on an embedded heterogamous dual-core platform. Therefore, this study develops and implements computer vision and sensing techniques of nighttime vehicle detection, collision warning determination, and traffic event recording. The proposed system processes the road-scene frames in front of the host car captured from CCD sensors mounted on the host vehicle. These vision-based sensing and processing technologies are integrated and implemented on an ARM-DSP heterogamous dual-core embedded platform. Peripheral devices, including image grabbing devices, communication modules, and other in-vehicle control devices, are also integrated to form an in-vehicle-embedded vision-based nighttime driver assistance and surveillance system.

INTRODUCTION

Traffic accidents have become a major cause of death. Most traffic accidents are caused by driver carelessness under traffic conditions. Therefore, detecting on-road traffic conditions for assisting drivers is a promising approach to help drivers take safe driving precautions. Accordingly, many studies have developed valuable driver assistance techniques for detecting and recognizing on-road traffic objects, including lane markings, vehicles, raindrops, and other obstacles. The objects are recognized from images of road environments outside the host car [1,2]. These driver assistance techniques are mostly developed based on camera-assisted systems, and can help drivers perceive possible dangers on the road or automatically control the apparatus of the vehicle (e.g., headlights and windshield wipers).

Recent studies on vision-based driver assistance systems [3–10] attempt to identify vehicles, obstacles, traffic signs, raindrops, pedestrians, and other patterns in on-road traffic scenes from captured image sequences using image processing and pattern recognition techniques. By adopting different concepts and definitions of interesting objects on the road, applying different techniques to capture image sequences can detect vehicles or obstacles. In previous studies on detecting vehicles in an image sequence for driving safety, the vehicle detection process is mostly conducted by searching for specific patterns in the images based on typical features of vehicles, such as edge features, shape templates, symmetrization, or vehicles' surrounding bounding boxes [3–10]. However, most studies on this topic have focused on detecting vehicles under daytime road environments.

At night, and under darkly illuminated conditions in general, headlights and taillights are often the only salient features of moving vehicles. In addition, many other sources of illumination that coexist with vehicle lights are found in road environments, including street lamps, traffic lights, and road reflector plates. These non-vehicle illumination objects make it difficult to obtain accurate cues for detecting vehicles in nighttime road scenes. To detect salient objects in nighttime traffic surveillance applications, Beymer *et al.* [11] applied a feature-based technique that extracts and tracks the corner features of moving vehicles instead of their entire regions. This approach works in both daytime and nighttime traffic environments, and is more robust to partial or complete occlusions. However, it suffers from high computational costs because it must simultaneously process numerous features of moving vehicles. Huang *et al.* [12] used block-based contrast analysis and inter-frame change information to detect salient objects at night. This contrast-based method can effectively detect outdoor objects in a surveillance area using a stationary camera. However, contrast and inter-frame change information are sensitive to the lighting effects of moving vehicle headlights, which often results in erroneous vehicle detection.

Recently, vehicle lights have been used as salient features for nighttime vehicle detection applications for driver assistance systems [13–15]. Stam *et al.* [13] developed an optical sensor array based system for detecting the vehicle lights. Based on an optical sensor array system set up on the vehicle, lighting objects appearing in the viewable area ahead of the host vehicle are imaged on the sensor array. Then a set of pre-determined thresholds are utilized to label bright-spotted pixels having gray intensities above the thresholds to determine the appearance of target vehicles. In Eichner and Breckon's headlight detection method [14], a rule-based lighting object detection approach is conducted by segmenting the lighting objects using a fixed threshold along with a headlight pairing rules, and then a temporal tracking is combined to refine the detection results. The above-mentioned techniques can detect the appearance of vehicles under a nighttime road environment with few lighting sources. However, because these techniques use a set of fixed threshold values which were configured beforehand, they are unable to adaptively adjust the selection of threshold values to match different nighttime lighting conditions. Therefore, their reliability in handling the circumstances where road environments have various lighting conditions, is limited. To improve the feasibility on nighttime vehicle light detection, O'Malley *et al.* [15] presented a taillight detection approach, which adopts the hue-saturation-value (HSV) color features for taillight detection and integrates a Kalman filter tracking process. By integrating the HSV color features, this approach can efficient detect taillights with red lighting characteristics under various road environments, including urban, rural, and motorway environments. However, the utilization of multiple color components of HSV color spaces requires additional computational costs on color transformation because of the floating-point computation of the HSV color features, and the Kalman filtering also suffers large computational costs on numerical and inverse matrix computations. Thus, this approach is inappropriate for implementing on the portable embedded systems with limited computational resources.

Most of the previous systems are implemented on personal computer (PC)-based platforms, and therefore lack the portability and flexibility required for installation in vehicular environments. The computational power and flexibility of embedded systems have recently increased significantly because of the development of system-on-chip technologies, and the advanced computing power of newly released multimedia devices. Thus, performing real-time vision-based vehicle and object detection in driver safety applications has become feasible on modern embedded platforms for driver assistance systems [16–18].

In current requirements for driver assistance systems, driving event data recording is also an important functionality for traffic accident reconstruction. Unfortunately, most previously developed systems do not integrate the driving event recording functions with the driver assistance functions into a stand-alone portable system. Efficiently recording the real-time image sequences of traffic conditions with driver actions is essential to meeting the demands of monitoring and recording driving event data. For this purpose, the transform coding technique is the most popular method for compressing monitoring image frames. As the fundamental development of this field, discrete cosine transform (DCT)-based coding has been commonly used, and has since become an element of the JPEG image compression standard. Accordingly, it has been applied to numerous electronic devices today. Researchers have recently demonstrated that discrete wavelet transform (DWT)-based transform coding outperforms DCT-based methods [19–28]. Hence, newly developed image coding methods, such as the still image compression standard JPEG2000 [25,26] and the video coding method standard MPEG-4 [27,28], adopt concepts based on DWT features. The lifting scheme [23] nearly halves the time required to perform DWT computations. Therefore, this scheme has been incorporated into DWT-based image compression techniques. For example, zero-tree coding methods [21–24] can achieve the most coding efficiency to both DWT and DCT-based transform coding techniques.

However, traffic event video recording based on these image and video coding techniques, in addition to vision-based function modules, on a portable embedded system still suffers from computational problems. This is primarily because of the limited computational resources of embedded platforms. Recording traffic event videos during the long driving times requires huge storage space, and the storage space on an embedded portable platform is limited and expensive. Based on the requirements of traffic accident reconstruction, the system must record critical event videos of potential accidents. Thus, the limited storage space on a portable embedded system can be used efficiently, and the responsibility of possible traffic accidents can be more effectively reconstructed and identified. In this manner, the proposed vision-based nighttime vehicle detection and driver warning approaches can enable the accurate and timely determination of possible traffic accidents. Therefore, the proposed system can efficiently activate the traffic event video recording process when possible traffic accidents might occur as a result of driver negligence or inappropriate driving behaviors. To provide a satisfactory solution for these issues, this study adopts a heterogamous dual-processor embedded system platform, and the traffic event video recording function and vision-based driver assistance modules, including vehicle detection, traffic condition analysis, and driver warning functions are implemented and optimized using the computational resources of the two heterogamous processors.

This study proposes a vision-based intelligentnighttime driver assistance and surveillance (VIDASS) system. The proposed VIDASS system includes the computer vision and sensing techniques of nighttime vehicle detection, collision warning determination, and traffic event recording functions by processing the road-scene frames in front of the host car, which are captured from the CCD sensors mounted inside the host vehicle. These proposed vision-based sensing and processing technologies are implemented as a set of embedded software component and modules based on a component-based system framework, and are integrated and performed on an ARM-DSP heterogamous dual-core embedded platform. Peripheral devices, including image grabbing devices, network communication modules, and other in-vehicle control devices, are also integrated to produce an in-vehicle embedded vision-based nighttime driver assistance and surveillance system. Accordingly, the goals and features of the proposed embedded driver assistance system are given as follows:

1. Effective detection and analysis road environment based on image segmentation, object recognition, and motion analysis;
2. Real-time event recording using efficient video compression and storage technology;
3. Configurable software framework to achieve the extension of the convenience and scalability;
4. A low-cost and high performance night driving assistance system implemented on a heterogamous dual-processor (ARM-DSP core) embedded system platform.

Experimental results show that the proposed system provides both efficiency and feasible advantages for integrated vehicle detection, collision warning, and traffic event recording for driver surveillance in various nighttime road environments and different traffic conditions.

THE PROPOSED NIGHTTIME DRIVER ASSISTANCE AND EVENT RECORDING SYSTEM

The proposed vision-based intelligentnighttime driver assistance and surveillance (VIDASS) system integrates effective vision-based sensing and processing modules, including nighttime vehicle detection, collision warning determination, and event recording functionalities. These functions are implemented to identify target vehicles in front of the host car, estimating their distances, determine possible collision accidents, and record traffic event videos. The real-time vision-based sensing and processing modules of the proposed VIDASS system include bright object segmentation, spatial clustering process, rule-based vehicle identification, vehicle distance estimation, and traffic event warning and control signaling machinery. The following subsections describe these features.Figure 1 shows a flow diagram of the proposed vision processing modules for the VIDASS system.

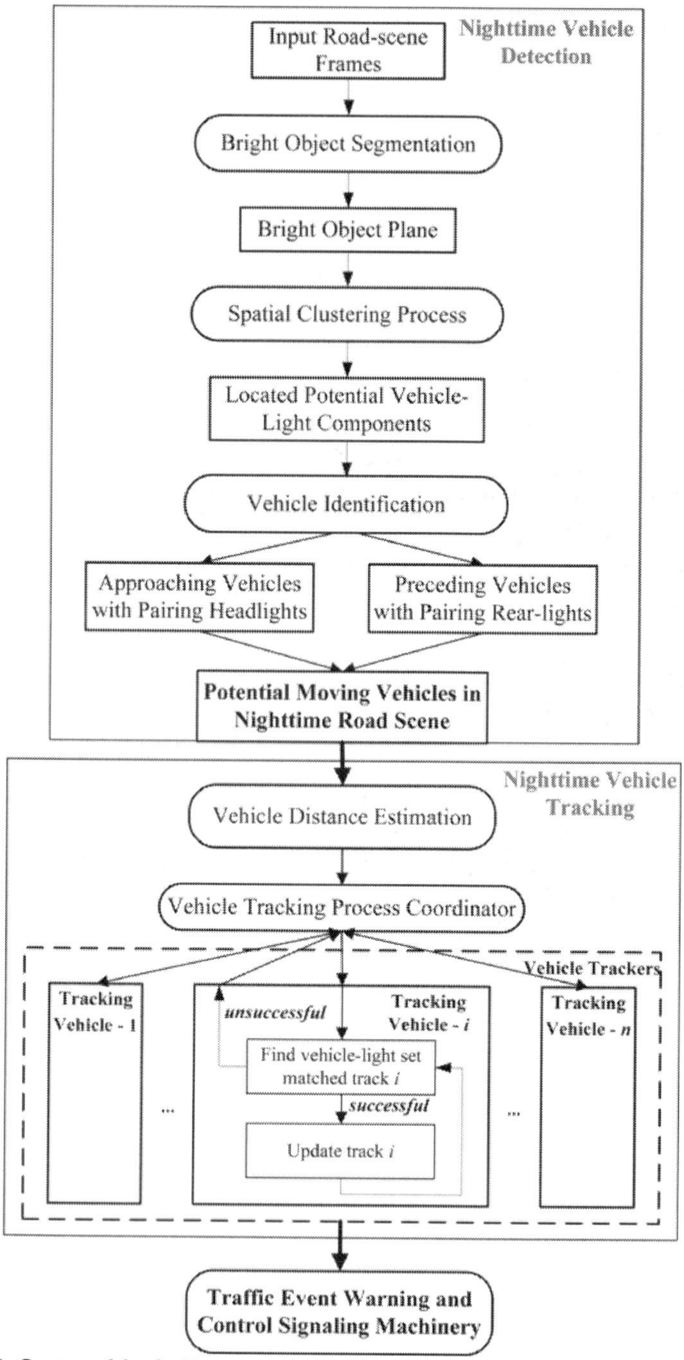

Figure 1. System block diagram.

Bright Object Segmentation Module

The input image sequences are captured from the vision system. These sensed frames reflect nighttime road environments appearing in front of the host car. Figure 2 shows a sample nighttime road scene taken from the vision system. In this sample scene, two vehicles are on the road. The left vehicle is approaching in the opposite direction on the neighboring lane, and the right vehicle is moving in the same direction as the camera-assisted host car.

Figure 2. A sample nighttime road scene.

The task of the bright object segmentation module is to extract bright objects from the road scene image to facilitate subsequent rule-based analysis. To reduce the computation cost of extracting bright objects, the module first extracts a grayscale image (Figure 3) (*i.e.*, the Y-channel) of the captured image by performing a RGB-to-Y transformation.

Figure 3. A sample grayscale nighttime road scene.

To extract these bright objects from a given transformed gray-intensity image, pixels of bright objects must be separated from other object pixels of different illuminations. Thus, an effective multilevel thresholding technique is required to automatically determine the appropriate number of thresholds for segmenting bright object regions from the road-scene image. For this purpose, we have proposed an effective automatic multilevel thresholding technique for fast region segmentation [29]. This technique can automatically decompose a captured road-scene image into a set of homogeneous thresholded images based on an optimal discriminant analysis concept. Extensive studies based on this optimal discriminant analysis concept have also been efficiently used in various object segmentation and analysis applications, such as text extraction for document image analysis [30], biofilm image segmentation [31,32], and multi-touch sensing detection applications [33].

Accordingly, to screen out non-vehicle illuminant objects such as street lamps and traffic lights located above half of the vertical y-axis (*i.e.*, the "horizon"), and save the computation cost, the bright object extraction process is only performed on the bright components located under the virtual horizon (Figure 4). Accordingly, as Figure 5 shows, after applying the bright object segmentation module on the sample image in Figure 2, pixels of bright objects are successfully separated into thresholded object planes under real illumination conditions.

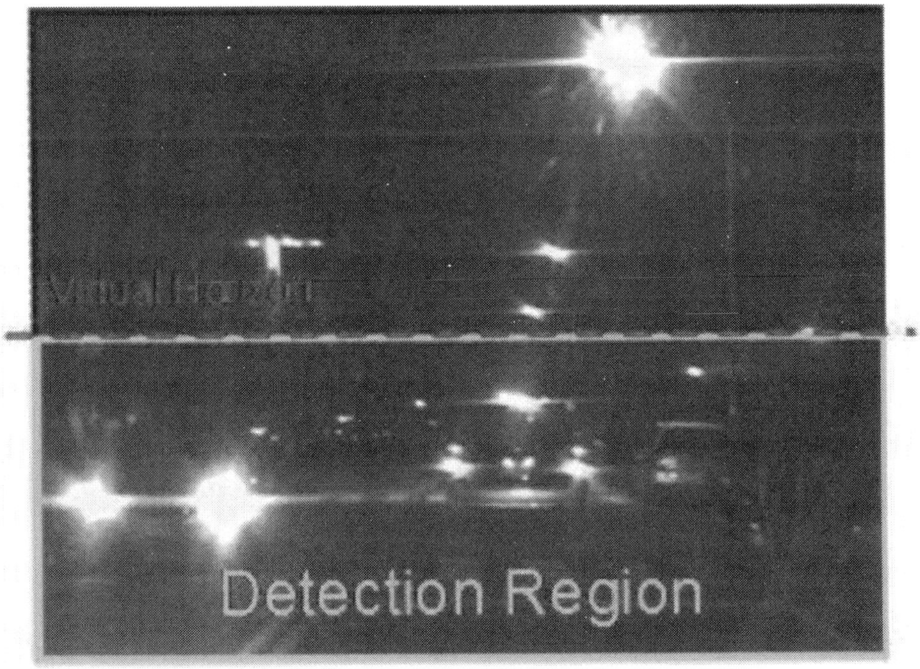

Figure 4. The detection region and virtual horizon for bright object extraction inFigure 2.

Figure 5. Results of performing the bright object segmentation module on the road scene images in Figure 2.

Spatial Analysis and Clustering Process Module

To identify potential vehicle-light components after performing bright object segmentation, a connected component extraction process is then performed on the bright object plane to locate the connected-components [34] of the bright objects. This process attempts to identify the horizontal-aligned vehicle lights; hence, a spatial clustering process is applied to the connected-components to cluster them into several meaningful groups. The resulting group includes a set of connected components, and may consist of vehicle lights, traffic lights, road signs, and other illuminated objects that frequently appear in nighttime road scenes. These groups are then processed by the vehicle light identification process to identify the actual moving vehicles. The following steps outline the proposed projection-based spatial clustering process:

1. C_i denotes one certain bright connected-component to be processed.
2. CG_k denotes a group of bright components, $CG_k = \{C_i, i = 0, 1, 2, \ldots, p\}$, and the total number of connected components contained in CG_k is denoted as $N_{cc}(CG_k)$.
3. The locations of the bounding boxes of a certain component C_i employed in the spatial clustering process are the top, bottom, left and right coordinates of these components, and these locations are denoted as $t(C_i)$, $b(C_i)$, $l(C_i)$, and $r(C_i)$, respectively.
4. The width and height of a bright component C_i are denoted as $W(C_i)$ and $H(C_i)$, respectively.
5. The horizontal distance D_h and vertical distance D_v between two bright components are defined as:

$$D_h(C_i, C_j) = \max[l(C_i), l(C_j)] - \min[r(C_i), r(C_j)] \quad (1)$$

$$D_v(C_i, C_j) = \max[t(C_i), t(C_j)] - \min[b(C_i), b(C_j)] \quad (2)$$

If the two bright components are overlapping in the horizontal or vertical direction, then the value of $D_h(C_i,C_j)$ or $D_v(C_i,C_j)$ is negative.

1. The degree of overlap between the vertical projections of the two bright components can be computed as:

$$P_v(C_i, C_j) = -D_v(C_i, C_j)/\min[H(C_i), H(C_j)] \quad (3)$$

To preliminarily screen out non-vehicle illuminant objects such as street lamps and traffic lights, first filter out the bright components located above one-third of the vertical y-axis (*i.e.*, only the bright components located under the constraint line), Thus, only the "virtual horizon" in Figure 4 is considered. This is because vehicles located at a distant place on the road become very small light "points", and "converge" into a virtual horizon.

To determine the moving directions of detected vehicles, it is necessary to identify the bright components at potential headlights and taillights before performing the respective analyses. The distinguishable characteristics of taillights are red illuminated lights. However, when the preceding vehicles are close to the camera-assisted car (*i.e.*, within 30 m), their taillights cause "blooming effects" in CCD cameras, and are usually too bright to appear as white objects in the captured images. As a result, only pixels located around the components of potential taillights have a distinguishable red appearance. Hence, the following *red-light criterion* is used to check if a bright component contains a potential taillight object. This criterion is determined as:

$$R_p(C_i) - T_{red} > \text{both } G_p(C_i) \text{ and } B_p(C_i) \tag{4}$$

where T_{red} is a predetermined threshold, and $R_p(C_i)$, $G_p(C_i)$, and $B_p(C_i)$ respectively represent the average intensities of the *R*, G, and*B* color frames of the pixels located at the peripheral of a given bright component C_i. The value of T_{red} is chosen as 10 to discriminate potential taillights and other bright components. If a bright component C_i satisfies the red-light criterion, then C_i is tagged as a *red-light*component; otherwise, it is tagged as a *non-red-light* component.

The connected components of bright objects are then recursively merged and clustered into bright component groups *CGs* if they have the same light tags, are horizontally close to each other, vertically overlapped, and aligned. In other words, if neighboring bright components satisfy the following conditions, they are merged with each other and clustered as the same group *CG*:

1. They have the same tags (*i.e.*, both are either *red-light* or *non-red-light* components).
2. They are horizontally proximate:

$$D_h(C_i, C_j) < T_d \times \max(H(C_i), H(C_j)) \quad (5)$$

3. They are highly overlapped in vertical projection profiles:

$$P_v(C_i, C_j) > T_p \quad (6)$$

4. They have similar heights:

$$H(C_S)/H(C_L) > T_h \quad (7)$$

where C_S is has a smaller height among the two bright components C_i and C_j, whereas C_L has a taller height.

In this study, T_d, T_p, and T_h represent the predetermined thresholds to determine the pairing characteristics of vehicle lights. These values are reasonably chosen as 3.0, 0.8, and 0.7, respectively. Figure 6 illustrates the processing results of the spatial clustering module applied to bright components. The spatial clustering process yields several groups of bright components, denoted as candidate vehicle-light groups.

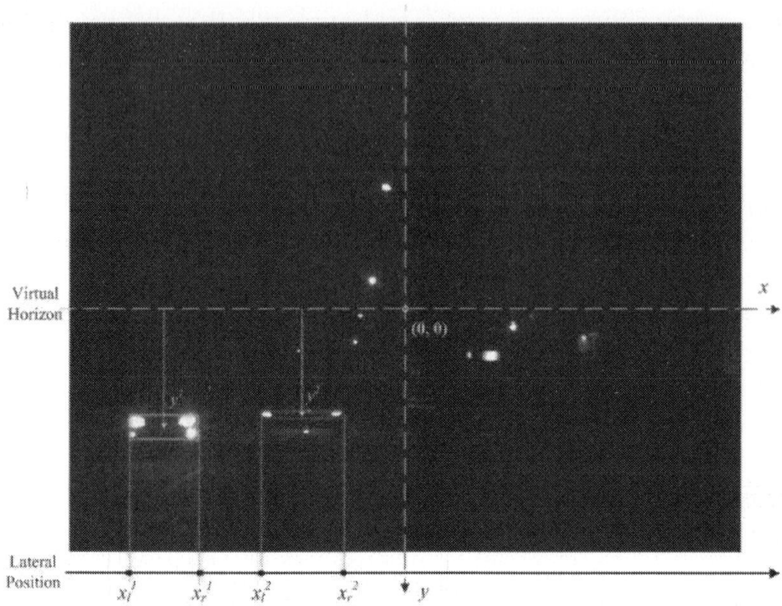

Figure 6. The spatial clustering process of the bright components of interest.

Vehicle Tracking and Identification Module

These methods obtain the light groups of potential vehicles in each captured frame. However, because sufficient features of some potential vehicles may not be immediately obtained from single image frames, a tracking procedure must be applied to analyze the information of potential vehicles from more successive image frames. The tracking information can then be used to refine the detection results and suppress occlusions caused by noise and errors during the bright object segmentation process and spatial clustering process. This tracking information can also be applied to determine useful information such as the direction of movement, positions, and relative velocities of potential vehicles entering the surveillance area.

The proposed vehicle tracking and identification module in the proposed vision system includes two phases. First, the phase of potential vehicle tracking associates the motion relation of potential vehicles in succeeding frames by analyzing their spatial and temporal features. The phase of vehicle recognition subsequently identifies actual vehicles among the tracked potential vehicles.

Potential Vehicle Tracking Phase

Considering only the detected object regions of potential vehicles in single frames may lead to the following problems in the segmentation process and the spatial clustering process: (1) a preceding vehicle may be too close to another vehicle moving parallel or street lamps, so that they may be occluded during the segmentation process, and thus, detected as one connected region; (2) an oncoming vehicle may be passing so close to the host car that it may be occluded by the reflected beams on the road, and hence be merged into one large connected region; and (3) the headlight set or taillight set of a vehicle may include multiple light pairs, and they may not be immediately merged into a single group by the spatial clustering process.

The motion of object regions of potential vehicles makes it possible to progressively process and refine the detection results of the potential vehicles by associating them in sequential frames. Therefore, this study presents a tracking process for potential vehicles that can effectively handle these problems. The tracking information of potential vehicles is also provided to the following vehicle identification and motion analysis processes to determine appropriate relative motion information regarding the target vehicles ahead of the host car.

When a potential vehicle is first detected in the field of view in front of the host car, a tracker is created to associate this potential vehicle with those in subsequent frames by applying spatial-temporal features. The following points describe the features used in this tracking process:

a. Pti denotes the ith potential vehicle appearing in front of the host car in frame t, and the bounding box, which encloses all the bright components contained in Pti, is denoted as $B(Pti)$.
b. The location of Pti employed in the tracking process is represented by its central position, and can be expressed as:

$$P_i^t = \left(\frac{l(B(P_i^t)) + r(B(P_i^t))}{2}, \frac{t(B(P_i^t)) + b(B(P_i^t))}{2} \right) \quad (8)$$

c. The overlapping score of the two potential vehicles *Pti* and *Pt'j*, detected at two different times *t* and *t'*, respectively, can be computed based on their area of intersection:

$$S_o(P_i^t, P_j^{t'}) = \frac{A\left(P_i^t \cap P_j^{t'}\right)}{Max\left(A(P_i^t), A(P_j^{t'})\right)} \quad (9)$$

d. The size-ratio feature of the enclosing bounding box of *Pti* is defined as:

$$R_s(P_i^t) = W(B(P_i^t)) \Big/ H(B(P_i^t)) \quad (10)$$

e. The symmetry score of the two potential vehicles *Pti* and *Pt'j* can then be obtained by:

$$S_s(P_i^t, P_j^{t'}) = R_s(P_S) \Big/ R_s(P_L) \quad (11)$$

where P_S is the one with the smaller size among the two potential vehicles *Pti* and *Pt'j*, and P_L is the larger one.

In each iteration of the tracking process for a newly incoming *frame t*, the potential vehicles appearing in the incoming frame, denoted as **P***t*={*Pti* | *i*=1,...,*k'*}, are analyzed and associated with the set of potential vehicles that have already been tracked in the previous frame *t* − 1, denoted as **TP***t*−1={*TPt*−1*j* | *j*=1,...,*k*}. The set of the tracked potential vehicles $\mathbf{TP}^t$ is subsequently updated according to the following process.

During the tracking process, a tracked potential vehicle might be in one of five possible states. The vehicle tracking process then applies different operations according to the given states of each tracked potential vehicle in each frame. The tracking states and associated operations for the tracked potential vehicles are as follows:

a. **Update**: When a potential vehicle *Pti*∈**P***t* in the current frame can match a tracked potential vehicle *TPt*−1*j*∈**TP***t*−1, then the tracker updates the set of the tracked potential vehicles $\mathbf{TP}^t$ by associating *Pti* with the tracker *TPtj* if the following matching condition is satisfied. The *matching condition* is defined as:
b.

$$\mathbf{m}\left(P_i^t, TP_j^{t-1}\right) > 0.6 \quad (12)$$

where the matching score **m**(*Pti*,*TPt*−1*j*) is computed as:

$$\mathbf{m}\left(P_i^t, TP_j^{t-1}\right) = w_o \cdot S_o\left(P_i^t, TP_j^{t-1}\right) + w_s \cdot S_s\left(P_i^t, TP_j^{t-1}\right) \quad (13)$$

where w_o and w_s represent the weights of the overlapping score and the size-ratio feature, and are set at 0.5 and 0.5, respectively.

c. **Appear**: If a newly identified potential vehicle *Pti*∈**P***t* cannot match any *TPt*−1*j*∈**TP***t*−1 in the previous time, then a new tracker for this potential vehicle is created and appended to the updated set $\mathbf{TP}^t$.
d. **Merge**: A potential vehicle *Pti* at the current frame matches multiple tracked regions of potential vehicles *TPt*−1*j*, *TPt*−1*j*+1…,*TPt*−1*j*+*n*. This situation may occur if the headlight set or taillight set of a single vehicle consists of multiple light pairs that were not clustered into a single group in the previous frame, but are correctly merged by the spatial clustering process in the current frame. Therefore, if *TPt*−1*j*, *TPt*−1*j*+1…, and *TPt*−1*j*+*n* all satisfy the *matching*

These discriminating rules were obtained by analyzing many experimental videos of real nighttime road environments in which vehicle lights appear in different shapes, sizes, directions, and distances. The values of the thresholds used for these discriminating rules were determined to yield good performance in most nighttime road environments.

Accordingly, the proposed system recognizes a tracked potential vehicle as an actual vehicle when these vehicle identification rules are satisfied. A tracked vehicle is no longer recognized as an actual vehicle when it cannot satisfy the vehicle identification criteria over a number of frames (three frames in general) or when it disappears from the field of view.

Vehicle Distance Estimation Module

For estimating the distance between the camera-assisted car and detected vehicles, the proposed module applies the perspective range estimation model of the CCD camera introduced in [35]. The origin of the virtual vehicle coordinate system appears at the central point of the camera lens. The *X* and *Y*-coordinate axes of the virtual vehicle coordinate are parallel to the *x* and *y*-coordinates of the captured images, and the *Z*-axis is placed along the optical axis and perpendicular to the plane formed by the *X* and *Y* axes. A target vehicle on the road at a distance *Z* in front of the host car projects to the image at the vertical coordinate *y*. Thus, the perspective single-camera range estimation model presented in [32] can be used to estimate the *Z*-distance in meters between the camera-assisted car and vehicle detected using the equation:

$$Z = k \cdot ((f \cdot H)/y) \tag{17}$$

where *k* is a given factor for converting from pixels to millimeters for the CCD camera mounted on the car at the height *H*, and *f* is focal length in meters.

Event-Driven Traffic Data Recording Subsystem

The proposed traffic event warning and recording subsystem is an automatic control process in the proposed system. This automatic control process includes a vehicle headlight control process and a traffic event video recording process. When any oncoming vehicles are detected, the headlight control process automatically switches the headlights to low beam, and then reverts to high beams once the detected vehicles leave the detection zone. The warning voice and traffic event video recording process are also activated to notify drivers to slowly decelerate when the distance from the detected preceding vehicles is too small, and to record a traffic event video using the MPEG4 video codec [27,28]. Figures 7 and 8 present flowcharts of the headlight control process and the traffic event video recording process, respectively.

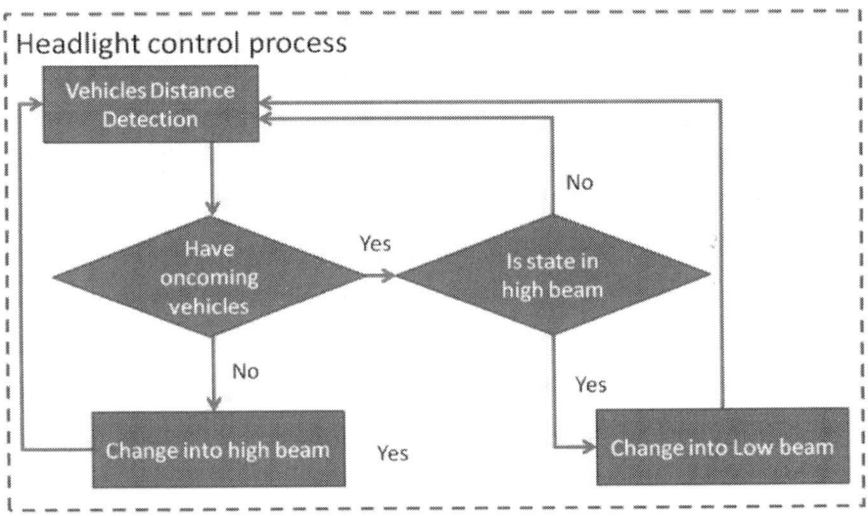

Figure 7. Headlight control process work flow.

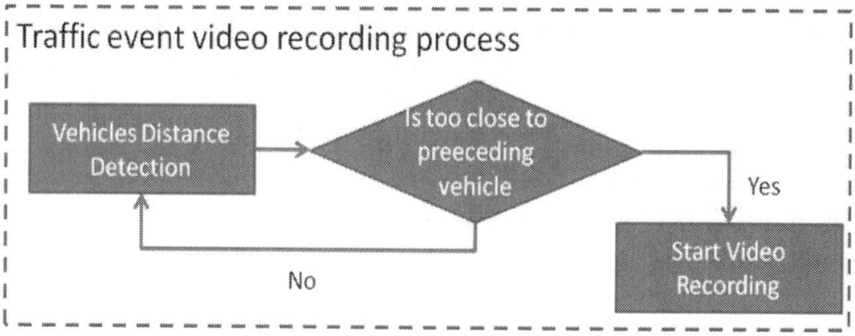

Figure 8. Traffic event video recording process work flow.

EMBEDDED SYSTEM IMPLEMENTATIONS

This section describes the implementation of the proposed VIDASS system. Subsection 3.1 presents the core architecture of the VIDASS system, including the integration of t with an heterogamous dual-core embedded platform with an embedded Linux OS and Qt graphical user interface (GUI) for user-friendly control machinery. This study also applies a component-based framework to implement the software framework of the proposed system. Subsection 3.2 describes the software architecture and implementation of the proposed VIDASS system.

Core Architecture

The proposed VIDASS system includes a set of vision-based modules and subsystems as described in the previous section, and these vision modules require an embedded portable platform that can provide efficient and optimized video-intensive computational resources and feasible embedded software development tools. For the aspects of hardware platform selection, the ARM general processors are famous for their low cost, low power consumption, extensive general-purpose IO control facilities, and sufficient OS awareness. However, the hardware architecture of ARM processors is designed to support general-purpose applications, and is therefore unsuitable for multimedia applications, such as image and video processing applications that cost large numerical computations on a large amount of multimedia data. In contrast, a digital signal processor (DSP), such as Texas Instruments (TI) C64x+ series, can provide optimized and accelerated computational instructions and units for video-intensive algorithms and multimedia data processing applications, although it lacks sufficient OS awareness compared with ARM processors. Thus, a complementary combination of the ARM general processor and the digital signal processor can provide a best-fitted solution for developing a vision-based system with the embedded OS, GUI module, and peripheral IO control facilities. In this way, we can divide the VIDASS system into a set of execution modules on the ARM and DSP processors according to their functional and computational properties to obtain effective performance and economic hardware costs.

In this study, the proposed VIDASS system was implemented on an ARM-DSP heterogamous dual-core embedded platform, the TI OMAP3530 platform (Figure 9) [36]. The TI OMAP3530 platform is an efficient solution for portable and handheld systems from the TI DaVinci™ Digital Video Processor family. This platform consists of one Cortex A8 ARM-based general-purpose processor with a 600 MHz operational speed and one TI C64x+ digital signal processor with a 430 MHz operational speed on a single chip. This platform also includes 256 MB of flash ROM memory for storing the embedded Linux OS kernel, Qt-based GUI system, and the proposed vision-based software modules, and 256 MB of DDR memory for executing the software modules. In addition to the main platform, peripheral devices such as image grabbing devices, LCD touchscreen panel, mobile communication module, and other in-vehicle control devices are also integrated to accomplish an in-vehicle embedded vision-based nighttime driver assistance and surveillance system.

Figure 9. The TI OMAP3530 Embedded Experimental Platform.

As for the implementation issue on the embedded platform with heterogamous processors, the processor architectures, instruction sets, and clock rates of the ARM and the DSP are different. Thus, the efficient system-level integration of these two heterogamous processors on multimedia application software remains a problem for technicians developing efficient solutions. From a developer's viewpoint, it is necessary to develop the main components of the system (such as the main application, GUI, and peripheral IO control facilities) on the ARM-side, in addition to the embedded operating system (OS) environments (such as the embedded Linux OS). Thus, it would be better to adopt the digital signal processor as a hardware computing resource in the embedded OS at the ARM-side, and assign the computational jobs of the desired vision-based algorithms and event video recording functions to be performed on the DSP-side. Transferring the multimedia data between the ARM and DSP can be achieved using the shared memory machinery and direct memory accesses. Based on these concepts, the TI provides a series of Linux Digital Video Software Development Kits (DVSDK) solutions for developing multimedia intensive application systems on ARM-DSP heterogamous dual-core platforms [37] of TI DaVinci™ Digital Video Processor family. The software architecture diagram provide in [37] shows the core architecture of the DVSDK. Thus, the proposed VIDASS system was implemented on a TI OMAP3530 platform with the DVSDK architecture. Based on this software development architecture, the proposed VIDASS system can be easily migrated, customized and extended on various embedded platforms of TI DaVinci™ Digital Video Processor family by the system developers for different demands of nighttime driver assistance applications.

Software Implementation

For the embedded software implementation on the OMAP3530 platform, the vision-based subsystem modules of the proposed VIDASS system (as described in the previous section) are implemented on the DSP-side to guarantee the real-time computational performance of image and vision processing modules. The main GUI system and peripheral IO control facilities were executed on the ARM-side along, in addition to the embedded Linux. The proposed VIDASS system uses the TI DVSDK software architecture to accomplish the real-time vision-based computing modules on the ARM-DSP heterogamous dual-core embedded platform, as Figure 10illustrates. The study adopts a component-based framework to develop the embedded software framework of the proposed VIDASS system. The VIDASS software framework includes the following embedded vision computing modules: the bright object segmentation (BOS) module, the spatial analysis and clustering process (SCP) module, the vehicle tracking and identification (VTI) module, the vehicle distance estimation (VDE) module, and the event-driven traffic data recording subsystem (ETDS). The advantages of using a component-based system include the ease of updating new function modules and extending the system to any subsequent developments of new vision-based assistance functional modules.

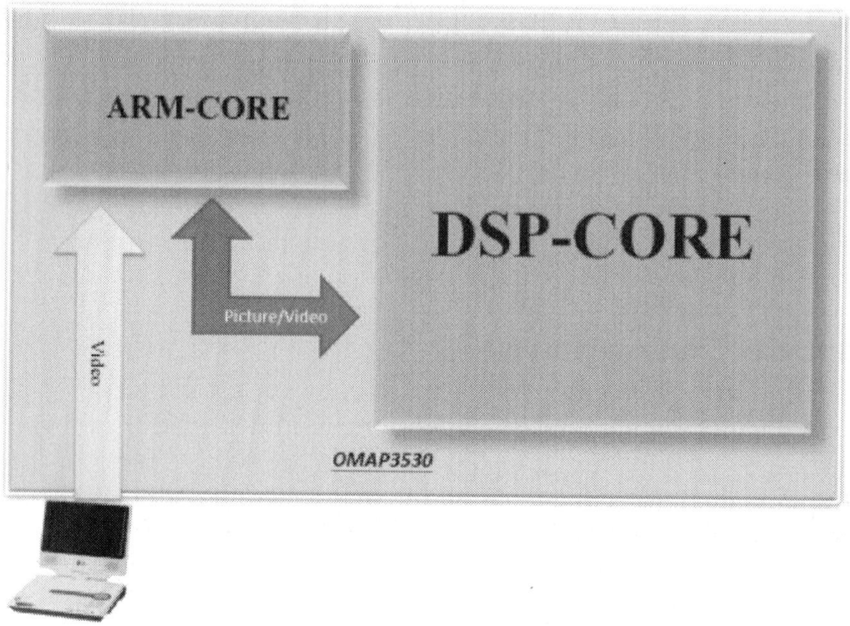

Figure 10. The software architecture of the VIDASS system on the ARM-DSP heterogamous dual-core embedded platform.

Figure 11 shows the software modules of the proposed VIDASS system. This figure shows that, based on the software modules, the proposed VIDASS system includes five processing stages: In the first stage, the BOS module obtains the grayscale image sequences captured from the vision system, and then applies the segmentation process to extract bright objects of interest.

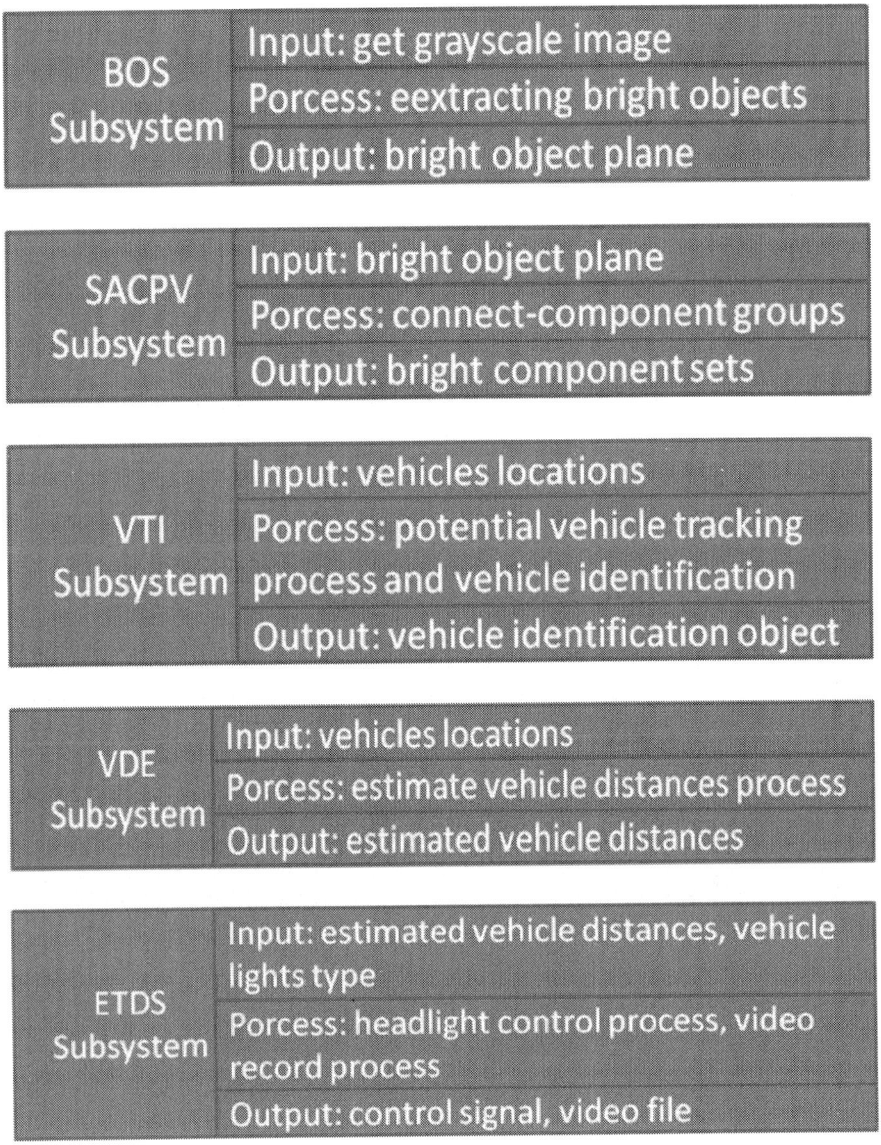

Figure 11. Software modules of the proposed VIDASS system.

The BOS module then exports the obtained bright object plane to the following stage. In the second stage, the SACPV module receives the bright object plane from the BOS module, applies a connected-component extraction process to group the bright objects into meaningful sets, and then sends out the resulting bright component sets. The third stage is the VIL module, which obtains the bright component sets from SACPV module and performs the rule-based identification process to detect and verify the oncoming and preceding vehicles. The VIL module then sends the detected vehicles and their locations out to the VDE module. In the fourth stage, the VDE module determines the relative distances between the host car and the detected oncoming and preceding vehicles, and then provides the estimated vehicle distances to the following stage. The final stage is the ETDS module, which performs and activates the headlight control process and activates the traffic event video recording according using the MPEG4 codec implemented by the DSP codec library [36,37] based the information of detected vehicles and their related distances.

The experiments in this study implemented these vision computing modules on a DSP-core to guarantee the computational efficiency of the optimized video processing instruction libraries provided by TI. To set the vision computing modules as a set of executive computation entities on the DSP-core, these vision modules are realized as the "customized codecs" according to the TI CodecGengine standard [38]. Thus, these modules can serve as the computational entities and are efficiently executed on the DSP-core based on the DVSDK software architecture [37]. Therefore, these vision computing modules can be effectively run on the DSP-core (Figure 12). The main system software executed on the ARM-core (Figure 13) can also efficiently apply and control these modules to obtain the processing results of the traffic conditions in front of the host car.

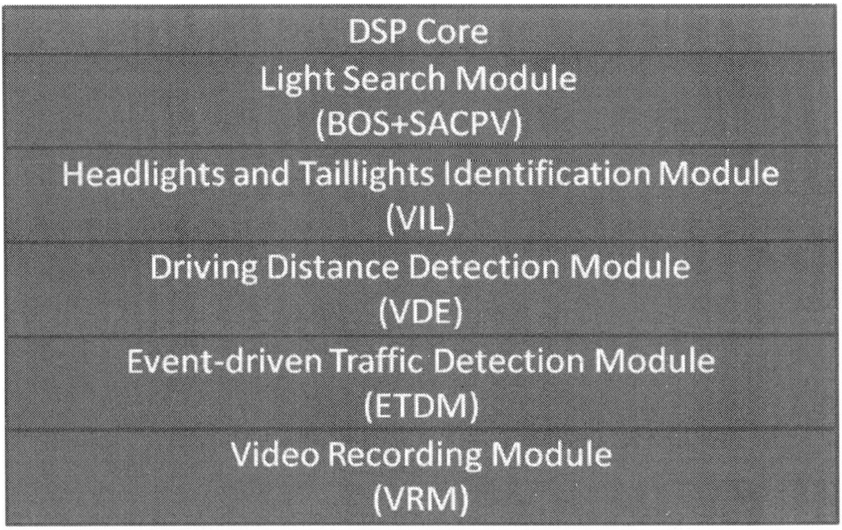

Figure 12. The software architecture of the vision computing modules on the DSP-core.

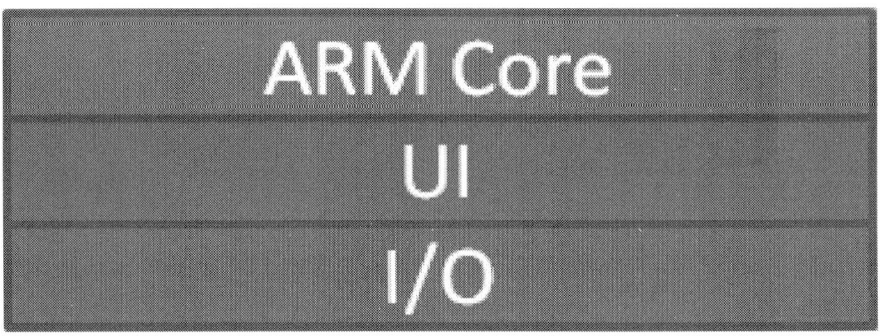

Figure 13. The software architecture of the main system on the ARM-core.

The main system software on the ARM-core was performed and integrated on an embedded Linux OS kernel, the Qt GUI SDK library. To control the hardware peripherals, the proposed system activates the corresponding control messages to the device drivers on the embedded Linux kernel. The Linux kernel then signals the related hardware peripherals, such as GPIO signals, voice speakers, and communication devices. Thus, drivers can operate the proposed system using the touch panel and hardware peripherals. Figure 14shows the overall software architecture of the proposed VIDASS system.

Figure 14. Proposed system software architecture.

EXPERIMENTAL RESULTS

The proposed vision-based techniques and modules of the VIDASS system were integrated and implemented on a TI IMAP3530 ARM-DSP dual-core embedded platform, which was set up on an experimental camera-assisted car. To make the proposed system operate well, the CCD camera for acquiring input image sequences of front road environments is mounted behind the windshield inside the experimental camera-assisted car with a sufficient height for capture a suitable region for covering the interesting driving lanes to be monitored, as shown in Figure 15. Then the view angle of the CCD camera is calibrated to be parallel to the ground for obtaining reliable features of target vehicles and the relative distances to them. In our experimental platform, the CCD camera is set at a 1.3 m height to the ground and parallel to the ground with an elevation angle of zero degree, while the focal length of the CCD camera is set as 10 mm. The peripheral devices, including image grabbing devices, mobile communication modules, and other in-vehicle control devices, were also integrated on the ARM-DSP dual-core embedded platform to accomplish an in-vehicle embedded vision-based nighttime driver assistance and surveillance system. The proposed vision system has a frame rate of 10 fps, and the size of each frame of captured image sequences is 720 pixels × 480 pixels per frame. The proposed system was tested on several videos of real nighttime road-scenes under various conditions.

Figure 15. Proposed system software architecture.

Figure 16 shows the main system screen of the proposed VIDASS system, which consists of three buttons, including the functions of the system configuration, system starting, and system stopping. The system configuration sets system values such as the voice volume, control signaling, and traffic event video recording, whereas system starting and system stopping buttons start and stop the proposed system, respectively.

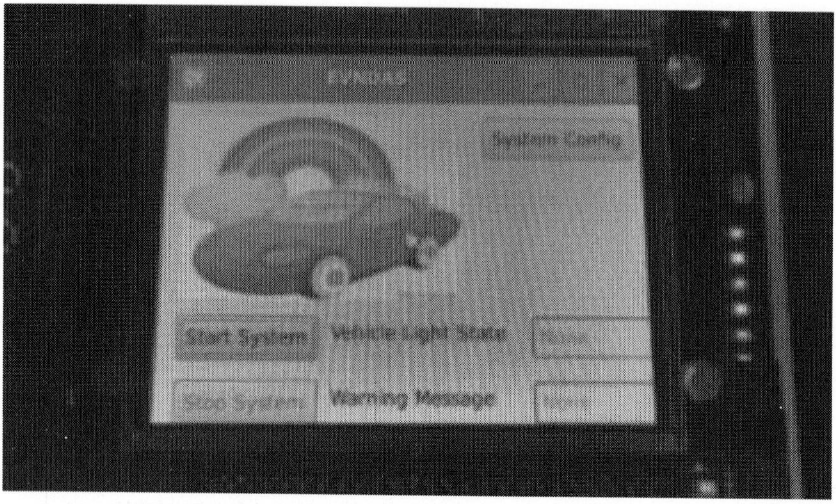

Figure 16. The user interface of the proposed VIDASS system.

The experiments in this study analyzed nine nighttime traffic video clips captured in urban, rural, and highway road environments with various traffic flows and ambient illumination conditions (test videos 1–9 in Figures 17–25).

Figure 17. Results of vehicle detection and event determination for a nighttime rural road under a normal illumination with free-flowing traffic conditions (Test video 1).

Figure 25. Results of vehicle detection and event determination for a nighttime highway under normal illumination and free-flowing traffic conditions (Test video 9).

The types of ambient lighting conditions in the experimental videos include normal illumination conditions, bright illumination conditions, and dim illuminations. Experimental videos in the normal illumination conditions were captured in the normal lit road environments, where the lighting sources are streetlights and vehicle lights. Experimental videos in the bright illumination conditions were captured in the highly lit urban road environments, where the lighting sources comprise of large amount of streetlights, building lights, neon lights and vehicle lights; while dim illumination videos were captured in unlit highway or rural road environments with only a few streetlights and vehicle lights. Each video lasted approximately 5–10 min. The locations of the detected preceding vehicles in the experimental samples are indicated by red rectangles, whereas the oncoming vehicles are labeled as green rectangles. When the host car comes too close to the detected preceding vehicles, the ETDS process warns the driver to slow to avoid collision dangers, and activates the traffic event recording process. When the proposed system detects oncoming vehicles, the ETDS module automatically switches the headlights from high beam to low beam.

First, the experimental videos in Figures 17–19 (Test videos 1–3) are evaluated at nighttime rural roads under normal and dim lighting conditions with free traffic flows. The snapshots in Figures 17–19 show that the proposed system correctly detects most of the preceding and oncoming vehicles, estimates their distances to the host car, and warns the driver to avoid possible collision dangers under different illumination conditions. The system also activates the event recording process when the host car drives too close to the target vehicles ahead.

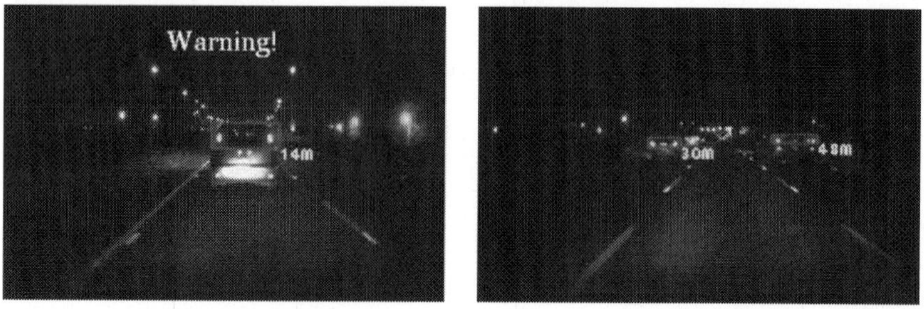

Figure 19. Results of vehicle detection and event determination for a nighttime rural road under a dim illumination with free-flowing traffic conditions (Test video 3).

Figures 20–22 (Test videos 4–6) depict experimental video scenes of urban roads under different illumination conditions with congested and free-flowing traffic conditions. As Figures 20–22 illustrate, the proposed system successfully detects the target oncoming and preceding vehicles with their distances to the host car, despite interference from numerous non-vehicle illumination objects in various urban road environments with different ambient lighting conditions. The system also activates warning and event recording processes under various environmental conditions associated with the complexity of road scenes.

Figure 20. Results of vehicle detection and event determination for a nighttime urban road under normal illumination and congested traffic conditions (Test video 4).

Figure 22. Results of vehicle detection and event determination for a nighttime urban road under bright illumination and free-flowing traffic conditions (Test video 6).

Figures 23–25 (Test videos 7–9) show further experimental videos of highway environments under dim and normal illumination conditions with different traffic conditions. The snapshots in these figures demonstrate that the proposed system can also correctly detect oncoming and preceding vehicles, estimate their distances, and activate warning and event recording processes. In these three samples, when the oncoming vehicles appear and are detected by the proposed system, the ETDS module switches the headlights to low beam to avoid dazzling oncoming drivers. The ETDS modules then switches the headlights back to high beam after the oncoming vehicles have passed the host car.

Figure 23. Results of vehicle detection and event determination for a nighttime highway under dim illumination and free-flowing traffic conditions (Test video 7).

To quantitatively evaluate vehicle detection performance, this study applies the Jaccard coefficient [39], which is commonly used to evaluate information retrieval performance. This measure is defined as:

$$J = \frac{T_p}{T_p + F_p + F_n} \quad (18)$$

where T_p (true positives) represents the number of correctly detected vehicles, F_p (false positives) represents the number of falsely detected vehicles, and T_n (false negatives) is the number of missed vehicles. The Jaccard coefficient J for the vehicle detection results of each frame of the traffic video sequences was calculated by manually counting the number of correctly detected vehicles, falsely detected vehicles, and missed vehicles in each frame. The average value of the Jaccard coefficients J, which serves as the detection score, was then obtained from all frames of the video sequences by:

$$\bar{J} = \sum_N J/N \quad (19)$$

where N is the total number of video frames. The ground-truth of detected vehicles was obtained by manual counting. In the nine experimental videos shown sequentially from Figures 17–25, the numbers of preceding vehicles and oncoming vehicles appearing in the view of the experimental host car were 203 and 130, respectively. Table 1 shows the quantitative experimental results of the proposed system on vehicle detection performance. The performance of vehicle detection is relatively promising because the overall detection score reaches 92.08% in real nighttime road environments under various lighting conditions and complex traffic scenes.

In addition to quantitatively evaluating the vehicle detection performance of the proposed system, this study evaluates the detection accuracy of possible collision warning events of the ETDS module. The proposed system issues timely warnings to the driver to slow to avoid collision dangers, and automatically activates the traffic event recording process. Table 2 shows the quantitative detection accuracy data of the possible collision warning events of the proposed system. This table shows that the proposed system can achieve high detection accuracy of warning condition determination. The average event detection accuracy reaches 93.3% for different nighttime road environments with various lighting conditions and traffic conditions. Accordingly, the high detection accuracy on warning events demonstrates that the proposed system enables timely collision avoidance, and efficient storage usage for traffic event data recording.

Table 1. Quantitative experimental data of the proposed system on detecting preceding and oncoming vehicles.

Test videos	Detection Score J of preceding vehicles	Detection Score J of oncoming vehicles
Test video 1 (Figure 17)	91.6%	100%
Test video 2 (Figure 18)	92.6%	N/A
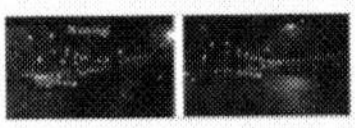 **Figure 18.** Results of vehicle detection and event determination for a nighttime rural road under a normal illumination with rainy and free-flowing traffic conditions (Test video 2). Click here to enlarge figure		
Test video 3 (Figure 19)	93.8%	N/A
Test video 4 (Figure 20)	90.9%	88.6%
Test video 5 (Figure 21)	93.0%	84.6%
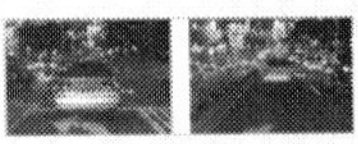 **Figure 21.** Results of vehicle detection and event determination for a nighttime urban road under bright illumination and congested traffic conditions (Test video 5). Click here to enlarge figure		
Test video 6 (Figure 22)	90.9%	92.6%
Test video 7 (Figure 23)	92.8%	82.5%
Test video 8 (Figure 24)	92.3%	83.1%

 **Figure 24.** Results of vehicle detection and event determination for a nighttime highway under normal illumination and slightly congested traffic conditions (Test video 8). Click here to enlarge figure		
Test video 9 (Figure 25)	94.4%	N/A
Total no. of preceding vehicles	203	
Total no. of oncoming vehicles	130	

Table 2. Detection accuracy of the proposed system on warning event determination.

Test videos	No. of actual warning conditions	No. of correctly determined warning conditions	Detection accuracy rate of warning events
Test video 1	8	7	87.5%
Test video 2	15	14	93.3%
Test video 3	17	16	94.1%
Test video 4	8	8	100%
Test video 5	19	17	89.4%
Test video 6	16	15	93.8%
Test video 7	10	9	90%
Test video 8	15	15	100%
Test video 9	12	11	91.7%
Total no. of actual warning conditions	120		
No. of correctly determined warning conditions	112		
Average event determination accuracy	93.3%		

For computational timing issues, the computation time required to process one input frame depends on the complexity of the traffic environment. Most of the computation time is spent on the spatial clustering process (including the connected component analysis) of potential lighting objects. Based on a TI OMAP3530 ARM-DSP embedded platform, the vision computing modules of the proposed VIDASS system require on average 70 milliseconds to process a video frame measuring 720 × 480 pixels, whereas the traffic event video recording takes approximately 10 ms per frame with hardware acceleration. The computational timing data of the vision processing modules are listed in Table 3. This computation cost ensures that the proposed system can effectively satisfy the demand of real-time processing at 12 fps (frames per second). Thus, the proposed system provides timely warnings and assistance for drivers to avoid possible traffic accidents.

Table 3. The computational timing data of the vision processing modules of the proposed VIDASS system.

Processing modules	Average computational times (ms per frame)
Bright object segmentation module	8.12
Spatial analysis and clustering process module	45.03
Vehicle tracking and identification module	12.41
Vehicle distance estimation module	3.77
Event-driven traffic data recording subsystem	10.36
Total computational times	79.7

CONCLUSIONS

This study presents a vision-based intelligentnighttime driver assistance and surveillance system (VIDASS system). The framework of the proposed system consists of a set of embedded software components and modules, and these components and modules are integrated into a component-based system on an embedded heterogamous multi-core platform. This study also proposes computer vision techniques for recording road-scene frames in front of the host car using a mounted CCD camera, and implement these vision-based techniques as embedded software components and modules. Vision-based technologies have been integrated and implemented on an ARM-DSP multi-core embedded platform and peripheral devices, including image grabbing devices and mobile communication modules. These and other in-vehicle control devices are also integrated to accomplish an in-vehicle embedded vision-based nighttime driver assistance and surveillance system. Experimental results demonstrate that the proposed system is effective and offers advantages for integrated vehicle detection, collision warning, and traffic event recording to improve driver surveillance in various nighttime road environments and traffic conditions. In the further studies, a feature-based vehicle tracking will be developed to more effectively detect and process the on-road vehicles having different number of vehicle lights, such as motorbikes. Besides, the proposed driver assistance system can also be improved and extended by integrating some sophisticated machine learning techniques, such as Support Vector Machine (SVM) classifiers, on multiple cues including vehicle lights and bodies, to further enhance the detection feasibility under difficult weather conditions (such as rainy and dusky conditions), and strengthen the classification capability on more comprehensive vehicle types, such as sedans, buses, trucks, lorries, small and heavy motorbikes.

This paper was supported by the National Science Council of Taiwan under Contract No. NSC-100-2219-E-027-006, NSC-100-2221-E-027-033, and NSC-100-2622-E-027-024-CC3. The authors would like to appreciate Ching-Chun Lin and Chian-Hao Wei for their helps on performing the experiments.

REFERENCES

1. Masaki, I. *Vision-Based Vehicle Guidance*; Springer-Verlag: New York, NY, USA; p. 1992.
2. Broggi, A.; Bertozzi, M.; Fascioli, A.; Conte, G. *Automatic Vehicle Guidance: The Experience of the ARGO Autonomous Vehicle*; World Scientific: Singapore; p. 1999.
3. Wu, B.-F.; Lin, C.-T.; Chen, Y.-L. Dynamic calibration and occlusion handling algorithms for lane tracking. *IEEE Trans. Ind. Electron* 2009, *56*, 1757–1773.
4. Broggi, A.; Bertozzi, M.; Fascioli, A.; Bianco, C.G.L.; Piazzi, A. Visual perception of obstacles and vehicles for platooning.*IEEE Trans. Intell. Transport. Syst* 2000, *1*, 164–176.
5. Jang, J.A.; Kim, H.S.; Cho, H.B. Smart roadside system for driver assistance and safety warnings: Framework and applications.*Sensors* 2011, *11*, 7420–7436.
6. Kurihata, H.; Takahashi, T.; Ide, I.; Mekada, Y.; Murase, H.; Tamatsu, Y.; Miyahara, T. Detection of raindrops on a windowshield from an in-vehicle video camera. *Int. J. Innov. Comp. Inf. Control* 2007, *3*, 1583–1591.
7. Musleh, B.; García, F.; Otamendi, J.; Armingol, J.M.; De la Escalera, A. Identifying and tracking pedestrians based on sensor fusion and motion stability predictions. *Sensors* 2010, *10*, 8028–8053.
8. Deguchi, D.; Doman, K.; Ide, I.; Murase, H. Improvement of a traffic sign detector by retrospective gathering of training samples from in-vehicle camera image sequences. Proceedings of the 10th Asian Conference Computer Vision, Queenstown, New Zealand, 8–12 November 2010; pp. 204–213.
9. Nedevschi, S.; Danescu, R.; Frentiu, D.; Marita, T.; Oniga, F.; Pocol, C.; Schmidt, R.; Graf, T. High accuracy stereo vision for far distance obstacle detection. *Proc. IEEE Intell. Veh. Symp* 2004. [CrossRef]
10. Betke, M.; Haritaoglu, E.; Davis, L.S. Real-time multiple vehicle detection and tracking from a moving vehicle. *Mach. Vision Appl* 2000, *12*, 69–83.
11. Beymer, D.; McLauchlan, P.; Coifman, B.; Malik, J. A real-time computer vision system for measuring traffic parameters. Proceedings of 1997 IEEE Computer Society Conference on Computer Vision and Pattern Recognition, San Juan, Puerto Rico, 17–19 June 1997; pp. 495–510.
12. Huang, K.; Wang, L.; Tan, T.; Maybank, S. A real-time object detecting and tracking system for outdoor night surveillance.*PatternRecognit* 2008, *41*, 432–444.

13. Stam, J.S.; Pierce, M.W.; Ockerse, H.C. Image Processing System to Control Vehicle Headlamps or Other Vehicle Equipment. U.S. Patent 6,868,322. 15 March 2005.
14. Eichner, M.L.; Breckon, T.P. Real-time video analysis for vehicle lights detection using temporal information. Proceedings of the 4th European Conference on Visual Media Production, London, UK, 27–28 November 2007; p. 1.
15. O'Malley, R.; Jones, E.; Glavin, M. Rear-lamp vehicle detection and tracking in low-exposure color video for night conditions.*IEEE Trans. Intell. Transp. Syst* 2010, *11*, 453–462.
16. Hsiao, P.-Y.; Yeh, C.-W.; Huang, S.-S.; Fu, L.-C. A portable vision-based real-time lane departure warning system: Day and night. *IEEE Trans. Veh. Technol* 2009, *58*, 2089–2094.
17. Hsiao, P.-Y.; Cheng, H.-C.; Huang, S.-S.; Fu, L.-C. CMOS image sensor with a built-in lane detector. *Sensors* 2009, *9*, 1722–1737.
18. Young, C.-P.; Chang, B.-R.; Tsai, H.-F.; Fang, R.-Y.; Lin, J.-J. Vehicle collision avoidance system using embedded hybrid intelligent prediction based on vision/GPS sensing. *Int. J. Innov. Comp. Inf. Control* 2009, *5*, 4453–4468.
19. Antonini, M.; Barlaud, M.; Mathieu, P.; Daubechies, I. Image coding using wavelet transform. *IEEE Trans. Image Proc* 1992,*1*, 205–220.
20. Li, S.; Li, W. Shape-adaptive discrete wavelet transforms for arbitrarily shaped visual object coding. *IEEE Trans. Circuits Syst. Video Technol* 2000, *10*, 725–743.
21. Shapiro, J.M. Embedded image coding using zerotrees of wavelets coefficients. *IEEE Trans. Sig. Process* 1993, *41*, 3445–3462.
22. Said, A.; Pearlman, W.A. A new, fast, and efficient image codec based on set partitioning in hierarchical trees. *IEEE Trans. Circuits Syst. Video Technol* 1996, *6*, 243–250.
23. Taubman, D. High performance image scalable image compression with EBCOT. *IEEE Trans. Image Process* 2000, *9*, 1158–1170.
24. Huang, H.-Y.; Wu, B.-F.; Chen, Y.-L.; Hsieh, C.-M. A new fast approach of single-pass perceptual embedded zero-tree coding. *ICIC Express Lett* 2011, *5*, 2523–2530.
25. *ISO/IEC 15444-1: Information Technology – JPEG2000 Image Coding System*; ISO/IEC: Geneva, Switzerland; p. 2000.
26. Christopoulos, C.; Skodras, A.; Ebrahimi, T. The JPEG2000 still image coding system: An overview. *IEEE Trans. Consumer Electron* 2000, *46*, 1103–1127.
27. Koenen, R. *Overview of the MPEG-4 Version 1 Standard; ISO/IEC JTC1/SC29/WG11 N1909*; ISO/IEC: Geneva, Switzerland, 1997.
28. Sikora, T. The MPEG-4 video standard verification model. *IEEE Trans. Circuits Syst. Video Technol* 1997, *7*, 19–31.

29. Wu, B.F.; Chen, Y.L.; Chiu, C.C. A discriminant analysis based recursive automatic thresholding approach for image segmentation. *IEICE Trans. Inf. Syst* 2005, *E88–D*, 1716–1723.
30. Chen, Y.L.; Wu, B.F. A multi-plane segmentation approach for text extraction from complex document images. *Pattern Recognit* 2009, *42*, 1419–1444.
31. Rueda, L. An efficient algorithm for optimal multilevel thresholding of irregularly sampled histograms. In *Structural, Syntactic, Statistical Pattern Recognition*; Springer-Verlag: Berlin, Germany, 2008; pp. 602–611.
32. Rojas, D.; Rueda, L.; Urrutia, H.; Ngom, A. Efficient optimal multi-level thresholding for biofilm image segmentation. In*Pattern Recognition in Bioinformatics*; Springer-Verlag: Berlin, Germany, 2009; pp. 307–318.
33. Chen, Y.-L.; Liang, W.-Y.; Chiang, C.-Y.; Hsieh, T.-J.; Lee, D.-C.; Yuan, S.-M.; Chang, Y.-L. Vision-based finger detection, tracking, and event identification techniques for multi-touch sensing and displaying systems. *Sensors* 2011, *11*, 6868–6892.
34. Suzuki, K.; Horiba, I.; Sugie, N. Linear-time connected-component labeling based on sequential local operations. *Comput. Vis. Image Underst* 2003, *89*, 1–23.
35. Stein, G.P.; Mano, O.; Shashua, A. Vision-based ACC with a single camera: Bounds on range and range rate accuracy. *Proc. IEEE Intell. Veh. Symp* 2003, 120–125.
36. The TI OMAP3530 Dual-Core Embedded Platform. Available online: http://www.ti.com/product/omap3530 (accessed on 9 January 2012)..
37. Linux Digital Video Software Development Kits (DVSDK). Available online: http://www.ti.com/tool/linuxdvsdk-dv (accessed on 9 January 2012)..
38. Codec Engine GenCodecPkg Wizard FAQ. Available online: http://processors.wiki.ti.com/index.php/Codec_Engine_GenCodecPkg_Wizard_FAQ (accessed on 9 January 2012)..
39. Sneath, P.; Sokal, R. *Numerical Taxonomy. The Principle and Practice of Numerical Classification*; W.H. Freeman: New York, NY, USA; p. 1973.

CITATION

Yen-Lin Chen, Hsin-Han Chiang, Chuan-Yen Chiang, Chuan-Ming Liu, Shyan-Ming Yuan and Jenq-Haur Wang, A Vision-Based Driver Nighttime Assistance and Surveillance System Based on Intelligent Image Sensing Techniques and a Heterogamous Dual-Core Embedded System Architecture, doi:10.3390/s120302373.

CHAPTER 7

FPGA-Based Real-Time Embedded System for RISS/GPS Integrated Navigation

Walid Farid Abdelfatah[1,], Jacques Georgy[1], Umar Iqbal[2] and Aboelmagd Noureldin[2,3]*

[1]Trusted Positioning Inc., Calgary, AB T2L 2K7, Canada; E-Mail:jgeorgy@trustedpositioning.com
[2]Electrical and Computer Engineering Department, Queen's University, Kingston, ON K7L 3N6, Canada; E-Mails: umar.iqbal@queensu.ca (U.I.);aboelmagd.noureldin@rmc.ca (A.N.)
[3]Electrical and Computer Engineering Department, Royal Military College of Canada, Kingston, ON K7K 7B4, Canada

ABSTRACT

Navigation algorithms integrating measurements from multi-sensor systems overcome the problems that arise from using GPS navigation systems in standalone mode. Algorithms which integrate the data from 2D low-cost reduced inertial sensor system (RISS), consisting of a gyroscope and an odometer or wheel encoders, along with a GPS receiver via a Kalman filter has proved to be worthy in providing a consistent and more reliable navigation solution compared to standalone GPS receivers. It has been also shown to be beneficial, especially in GPS-denied environments such as urban canyons and tunnels. The main objective of this paper is to narrow the idea-to-implementation gap that follows the algorithm development by realizing a low-cost real-time embedded navigation system capable of computing the data-fused positioning solution. The role of the developed system is to synchronize the measurements from the three sensors, relative to the pulse per second signal generated from the GPS, after which the navigation algorithm is applied to the synchronized measurements to compute the navigation solution in real-time. Employing a customizable soft-core processor on an FPGA in the kernel of the navigation system, provided the flexibility for communicating with the various sensors and the computation capability required by the Kalman filter integration algorithm.

INTRODUCTION

Navigation

Navigation is the science which comprises the methods and technologies to determine the time varying position, velocity and attitude of a moving object by utilizing either sensor-based or satellite-based measurements or by integrating the measurements from both navigation systems [1].

The Global Positioning System (GPS) is a satellite-based, absolute-positioning navigation system, developed by the US Department of Defense (DoD) in the early 1970s, used to provide time, position and velocity information [2]. Although, the navigation solution that is provided by GPS is sufficiently accurate, in the order of meters to centimeters, especially when augmented with other satellite-based or ground-based augmentation systems, it is unable to fulfill the requirements of continuity and reliability in some situations. As a satellite-based navigation system, GPS requires a line-of-sight (LOS) between the receiver's antenna and the satellites. In urban canyons, tunnels, and other GPS-denied environments, the LOS requirement can't be always met, resulting in GPS outages caused by GPS signal blockage, interference, jamming and multi-path effects. Signal interruption is one of the main reasons that affects the continuity and reliability of the GPS navigation solution. Due to the mentioned reasons, GPS can't provide a continuous and reliable solution when used as a stand-alone navigation system, and a better navigation solution can be obtained by integrating the measurements from one or more sensor-based systems, with GPS measurements.

The Inertial Navigation System (INS) is a sensor-based, self-contained, dead-reckoning navigation system in which measurements at a high sample rate, provided by an inertial measurement unit (IMU) are used by a navigation processor to compute the position, velocity and attitude of the moving object relative to a known starting position, velocity and attitude [3]. The IMU consists of a triad of accelerometers and a triad of gyroscopes. Measurements are integrated twice for accelerometers and once for gyroscopes to yield position and attitude. The calculated navigation states (i.e., position, velocity and attitude) drift with time due to the needed integrations and the sensors errors (such as biases, scale factors, and noise) which lead to unbounded accumulation of errors. Therefore, inertial sensors alone are unsuitable for accurate positioning over an extended period of time. The advantages and disadvantages of INS show that its characteristics are complementary with those of absolute positioning systems, such as GPS. Thus, proper integration of the measurements from both INS and GPS can provide great value by mitigating each other problems. The development of Micro-Electro-Mechanical Systems (MEMS) [4] based inertial sensors, made the utilization of MEMS-based INS for consumer applications more affordable as they can be provided at a lower cost. Furthermore they are small in size, light weight, and have low power consumption. However, the errors present in the inertial sensors, materialize by a greater magnitude when using MEMS-based inertial sensors.

Instead of integrating the GPS with a full IMU [5], containing three accelerometers and three gyroscopes, a reduced inertial sensor system (RISS) which utilizes only one gyroscope and an odometer or wheel encoders is integrated with GPS to provide a 2D positioning solution [6,7]. With the assumption that the vehicle stays in the horizontal plane most of the time, the vehicle speed derived from the odometer is used together with the heading information obtained from the gyroscope to determine the velocities along the East and North directions, from which the vehicles' longitude and latitude are determined. The position, velocity and heading errors are estimated by a Kalman filter (KF) relying on RISS dynamic error model and GPS position and velocity updates [6]. This method requires a dynamic error model for RISS, a measurement model for GPS updates, and a stochastic model of the gyroscope sensor error. Since GPS has a relatively consistent, long-term accuracy, it is used to update both RISS position and velocity components and thus prevent the long-term growth of the RISS errors. On the other hand, the accurate short-term information provided by the RISS is used to overcome GPS outages and multi-path errors. In case a GPS outage occurs, KF operates in prediction mode only, correcting the RISS information based on the system error model. More discussion on the land-based navigation systems, vehicular dead-reckoning and their errors is presented in [8–12].

Embedded Systems

Embedded systems are data processing systems that are embedded into larger systems (such as telecommunication equipment, transportation systems and consumer electronics) and not normally visible to the user [13]. Most embedded devices are designed to perform one dedicated function [14] which requires collecting data from the surrounding environment via a set of sensors, then manipulating the data according to a specific algorithm resulting in useful information for the users. Embedded systems are different than general purpose computer systems in many aspects. These systems differ in that they are required to be dependable, efficient and meet a set of real-time constraints derived from the system to be designed in order not to jeopardize the quality to be provided by the system [13–18]. Another characteristic of embedded systems is its method of software development, known as cross-platform development, where the embedded system's (i.e., target) software is developed on a host system which employs a general-purpose processor [18]. There exist several hardware platforms that can be used for developing an embedded system such as microcontrollers (μC), digital signal processors (DSP), field programmable gate arrays (FPGA) and application specific integrated circuits (ASIC) [17,19–22]. Choosing one platform over the other depends not only on the requirements of the system to be developed such as the performance, power consumption, cost per chip, but also on the ease of the tools accompanied by a specific platform to assist the developers in producing the system within the constrained system cost and project time.

The ultimate goal, after which a navigation algorithm is developed, is the realization of the algorithm on an embedded system where the measurements from the different sensors are acquired and synchronized, after which the navigation algorithm integrating these time-aligned measurements can be applied to output a real-time solution at a defined rate. The transition from the algorithm research to realization is a crucial step in assessing the practicality and effectiveness of the navigation algorithm and consequently the developed embedded system, either for a proof-of-concept or to be accepted as a product. In the transition process, there is no unique methodology that can be followed to design an embedded system. Depending on the platform of choice, different methodologies are used by the system designers to produce the final product. For example, designers which use processor-based cores such as microcontrollers, or digital signal processors make use of embedded software oriented methodologies to develop their systems, as their main design is revolving around firmware development rather than hardware. However, designers which use FPGAs as their development platforms, have the flexibility to use either the processor-based approach, developing their system entirely in firmware or hardware-based approach, developing their systems entirely in hardware. In other designs in which FPGAs are used, the overall system can be partitioned into hardware and firmware.

The main contribution of this paper is the implementation of an embedded real-time navigation system based on the 2D RISS/GPS navigation algorithm.

D-RISS/GPS INTEGRATED NAVIGATION SYSTEM

The RISS/GPS algorithm is an integrated navigation algorithm that integrates GPS with self-contained sensors that, unlike GPS, does not suffer from blockages which lead to the interruption of the positioning solution or multipath that leads to the degradation of the positioning solution. GPS helps calibrate the errors in the self-contained sensors that otherwise can cause the positioning error to grow unbounded. As self-contained sensors, the presented navigation algorithm uses a gyroscope whose axis is aligned with the vertical axis of the body of the moving platform and the platform speed readings from either wheel encoders (in case of the mobile robot) or from vehicle odometer (in case of a full sized land vehicle). The benefit of this integrated system for land-based navigation over the traditional INS/GPS integration that utilizes a full IMU (inertial measurement unit) is: (i) the elimination of the two horizontally aligned gyroscopes of the full IMU; and (ii) the use of the vehicle speed to derive velocity directly instead of using accelerometers to derive velocity in the full IMU setting. The first benefit mentioned above is because, during a GPS outage, in a full IMU setting the positioning error is proportional to the cube of the outage duration and the residual uncompensated horizontal gyroscopes bias drifts after the compensation for these biases by the state estimation technique used (such as KF). This is the worst and most influential error in traditional INS/GPS solutions during outages. Taking advantage of the fact that land vehicles move mostly in the horizontal plane and that any off-plane motion is extremely small compared to the in-plane motion, these two gyroscopes and their errors are totally removed from the presented system. The second advantage of the presented system is the use of the vehicle speed readings (from wheel encoders or odometer) together with non-holonomic constraints on land vehicles to derive the velocities instead of relying on accelerometers to derive velocity. Calculating velocity from accelerometers involve a mathematical integration operation, while their calculation form speed readings and heading does not involve such operation. The problem with mathematical integration operations is that it will cause any bias error to lead to a growing unbounded error with time. Using accelerometers, the error in velocity will be proportional to the outage duration and the residual uncompensated accelerometers bias drifts after the compensation for these biases by the state estimation technique used (such as KF); while the position error will be proportional to the square of duration and the uncompensated biases. So the presented RISS/GPS navigation solution will be better than the traditional solution an order of magnitude in duration of the outage because of the elimination of the first integration operation.

RISS Mechanization

RISS Mechanization [6,7] is the process involved in transforming the measurements of a RISS system acquired in the body frame (i.e., vehicle frame) into position, velocity and attitude. It is a recursive process based on the initial conditions, or the previous output and the new measurements.

For a vehicle moving in a 2D plane, five navigation states are computed consisting of two position parameters, two velocity parameters and one orientation parameter. The 2D RISS system readings ($\boldsymbol{\omega}_z$ for the gyroscope and $\mathbf{V}_{forward}$ for the odometer) are given in a coordinate frame known as the body frame (b-frame) while the position is expressed in the earth-centered, earth-fixed (ECEF) geodetic coordinate frame and the velocity and attitude are usually expressed in another coordinate frame known as the local-level frame. The local-level frame coordinates are formed from a plane tangent to the Earth's surface fixed to the current location of the moving object and hence it is sometimes known as a "Local Tangent" or "Local Geodetic" plane.

In 2D, the position parameters are the latitude ($\boldsymbol{\varphi}$) and the longitude ($\boldsymbol{\lambda}$), the velocity parameters are the East ($\mathbf{V}_{East}$) and North ($\mathbf{V}_{North}$) velocities, and the only orientation parameter is the azimuth (**A**). Figure 1 shows the measurements of the 2D-RISS system in the body frame, with respect to the local-level frame in which the results are expressed.

where:

- λ = Vehicle's Longitude (radians)
- R_M = Meridian radius of curvature of the Earth's ellipsoid (meters).

RISS/GPS Data Fusion Using KF

The conventional KF method is used, in a closed-loop, loosely coupled fashion, to fuse the position and velocity measurements from the GPS with the RISS computed position and velocity. KF [5,12,23] is a state estimation technique that can integrate measurement data to obtain state estimate by recognizing that the measurements are noisy, and might have only a small effect on the state estimate, or in some cases to be neglected completely. KF smoothes out the effects of noise in the state variables estimated, by incorporating more information from reliable data than unreliable data. If mechanization was run in an unaided fashion, the errors in its output are passed from one iteration to another thus leading for the navigation states errors to drift with time. Therefore, error models are required for the analysis and estimation of different error sources associated with the RISS system. Since the errors in dynamic systems are variable in time, they are described by differential equations. Linearization of a non-linear dynamic system is the most common approach to derive a set of linear differential equations that defines the error states of a dynamic system. Figure 2shows where the KF fits within the navigation solution. One of the main advantages of the KF, computation-wise, is that it is very convenient for real-time processing.

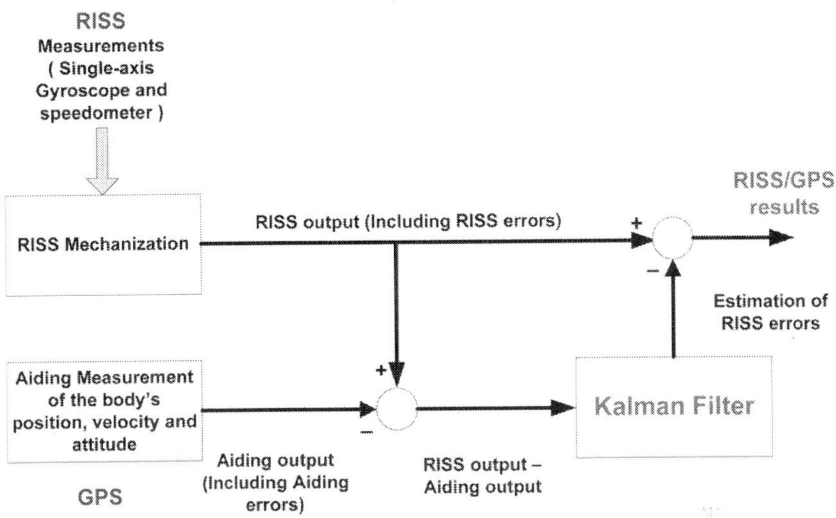

Figure 2. RISS/GPS Integrated Navigation Solution.

KF addresses the general problem of trying to estimate the state of a discrete-time process that is governed by the models:

1. State equation (System model or Process model):

$$\mathbf{x}_{k+1} = F_{k+1,k}.\mathbf{x}_k + G_k\mathbf{w}_k \tag{4}$$

where:

- $\mathbf{x}$ = Error state vector
- F = State transition matrix
- G = Noise coupling matrix
- $\mathbf{w}$ = System/Process noise.

2. Observation equation (Measurement model):

$$\mathbf{z}_{k+1} = H_{k+1} . \mathbf{x}_{k+1} + \boldsymbol{\nu}_{k+1} \quad (5)$$

where:

- **z** = Observation vector (i.e., Measurement vector)
- H = Measurement Design matrix
- **v** = Measurement noise.

The corresponding error-state system model used by KF is presented in Equation (6)where $\delta\mathbf{x}_{k+1}$ is the error-state vector, $a_{forward}$ is the acceleration derived from odometer data, T_s is the sampling time, w_k is a unit variance white Gaussian noise. The error-state vector has 7 error states. These are the latitude and longitude errors ($\delta\phi$, $\delta\lambda$), the East and North velocities errors (δV_{East}, δV_{Noth}), the azimuth error (δA), the stochastic gyroscope error ($\delta\omega z$) and the residual errors associated with the odometer-derived acceleration ($\delta a_{forward}$). The stochastic errors associated with the gyroscope and the odometer-derived acceleration are modeled by Gauss-Markov models, $\gamma_{aforward}$ is the inverse of the autocorrelation time for odometer-derived acceleration noise, σ2aforward is the variance of odometer-derived acceleration noise, $\beta_{\omega z}$ is the inverse of the autocorrelation time for gyroscope noise, σ2ωz is the variance of gyroscope noise.

$$\underbrace{\delta \mathbf{x}_k = \begin{bmatrix} \delta\varphi \\ \delta\lambda \\ \delta V_{East} \\ \delta V_{North} \\ \delta A \\ \delta a_{forward} \\ \delta\omega_z \end{bmatrix}}_{k+1}$$

$$= \left[I_{7\times 7} + \begin{pmatrix} 0 & 0 & 0 & \frac{1}{R_M+h} & 0 & 0 & 0 \\ 0 & 0 & \frac{1}{(R_N+h)\cos\varphi} & 0 & 0 & 0 & 0 \\ 0 & 0 & 0 & 0 & a_{forward}\cos A & \sin A & 0 \\ 0 & 0 & 0 & 0 & -a_{forward}\sin A & \cos A & 0 \\ 0 & 0 & 0 & 0 & 0 & 0 & 1 \\ 0 & 0 & 0 & 0 & 0 & -\gamma_{a_{forward}} & 0 \\ 0 & 0 & 0 & 0 & 0 & 0 & -\beta_{\omega_z} \end{pmatrix} T_s \right]$$

$$\underbrace{\begin{bmatrix} \delta\varphi \\ \delta\lambda \\ \delta V_{East} \\ \delta V_{North} \\ \delta A \\ \delta a_{forward} \\ \delta\omega_z \end{bmatrix}}_{k} + \begin{bmatrix} 0 \\ 0 \\ 0 \\ 0 \\ 0 \\ \sqrt{2\gamma_{a_{forward}}\sigma^2_{a_{forward}}} \\ \sqrt{2\beta_{\omega_z}\sigma^2_{\omega_z}} \end{bmatrix} w_k \quad (6)$$

Since loosely coupled integration is used, GPS position and velocity updates are used during the update stage of KF. The measurement model used by KF is:

$$\delta \mathbf{z}_k = \begin{bmatrix} \varphi^{RISS} & - & \varphi^{GPS} \\ \lambda^{RISS} & - & \lambda^{GPS} \\ V_{East}^{RISS} & - & V_{East}^{GPS} \\ V_{North}^{RISS} & - & V_{North}^{GPS} \end{bmatrix}_{k+1} = \begin{bmatrix} 1 & 0 & 0 & 0 & 0 & 0 & 0 \\ 0 & 1 & 0 & 0 & 0 & 0 & 0 \\ 0 & 0 & 1 & 0 & 0 & 0 & 0 \\ 0 & 0 & 0 & 1 & 0 & 0 & 0 \end{bmatrix} \begin{bmatrix} \delta\varphi \\ \delta\lambda \\ \delta V_{East} \\ \delta V_{North} \\ \delta A \\ \delta a_{forward} \\ \delta\omega_z \end{bmatrix}_{k+1} + \delta \boldsymbol{\nu}_{k+1} \quad (7)$$

EMBEDDED SYSTEM DESIGN

Due to the complexity of the embedded systems, design methodologies are needed to guide the decisions of the designers when developing large systems. The selected design process, shown in Figure 3, also defines the constraints between different teams that are working on the system. The design methodology acts as a framework that will direct the work flow and organize the tasks in relation to the different phases in the system life cycle. It has to be noted that, unlike the waterfall model in which the flow of information and work from higher levels of abstraction (i.e., earlier phases) to more detailed steps (i.e., later phases) involves a limited amount of feedback to the upper higher levels of abstraction, the work focus can move from one phase to another in the backward direction allowing system refinement and optimization according to the experiences gained through the work [9,10].

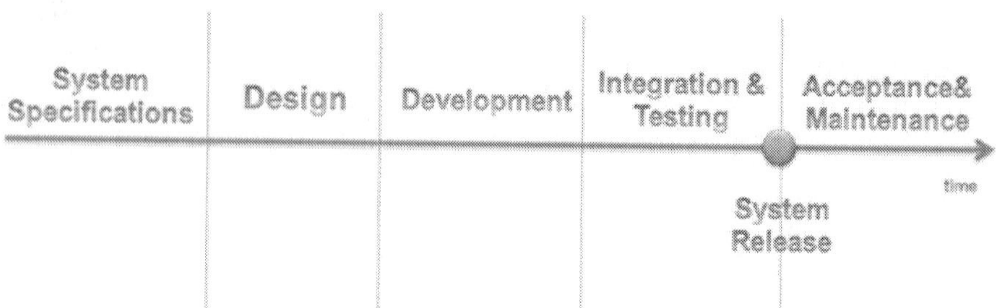

Figure 3. Embedded System Design Life Cycle.

System Requirements and Specifications Definition Phase

The process of designing an embedded system starts with the definition of the system—the goal of the research work. The system definition describes what the embedded system is to be and do. The functional requirements for the real-time navigation embedded system are as follows:

1. Interface with the NovAtel SPAN unit [24] which includes the OEM4 GPS receiver [25], Crossbow inertial measurement unit (IMU) IMU300CC-100 (the Crossbow IMU is a MEMS grade low-end device that uses three orthogonal accelerometers and gyroscopes. However, the 2D RISS/GPS algorithm uses the

vertical gyroscope only within the IMU which is aligned with the vehicle's z-axis) [26] and a mobile robot's encoders or the vehicle's odometer via the ElmScan vehicle speed scan tool [27], using the universal asynchronous receiver transmitter (UART) serial communication interface. For each sensor, a UART channel is required.

2. Synchronize the data from the three sensors using the pulse per second (PPS) signal from the GPS, for a more reliable fusion solution.
3. Extract the information required from these sensors to get the position, velocity and heading from the GPS receiver; the angular rotation rate from the vertical gyroscope within the IMU which is aligned with the vehicle's z-axis; and the vehicle's speed from the wheel encoders or the odometer.
4. Once the synchronized extracted information is available, the KF for 2D RISS/GPS navigation algorithm is applied, to yield a consistent navigation solution per second. The rate of the navigation solution is 1 Hz due to utilizing the robot's wheel encoders or the vehicle's odometer in the system, where the speed readings are acquired at 1 Hz relative to the PPS synchronization signal. The navigation solution has to be computed and available before the new incoming GPS PPS synchronizer signal and measurements from the three sensors are available as shown in Figure 4.

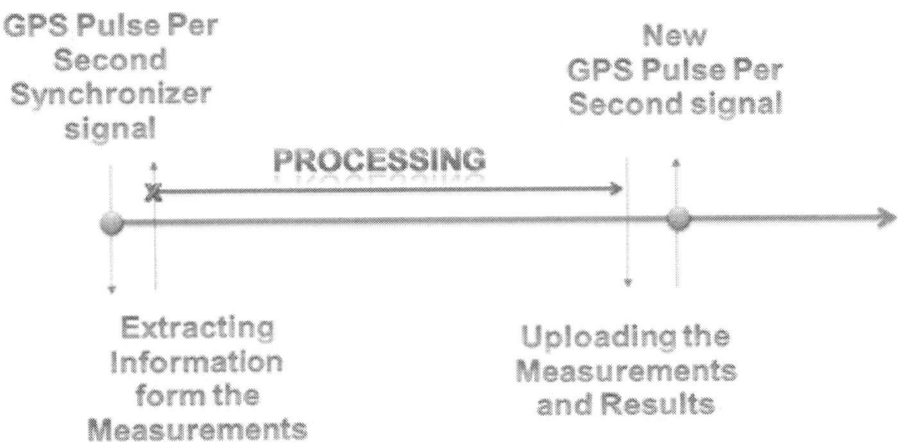

Figure 4. Measurements Synchronization, Information Extraction and Processing at 1 Hz rate.

It has to be noted that the navigation solution computed by the RISS/GPS algorithm is bounded by the rate of system sensors. The solution computed by the developed real-time system is at a rate of 1 Hz due to the speed reading which is acquired from the robot's wheel encoders or the vehicle's odometer at a 1 Hz rate with respect to the arrival of the PPS signal from the GPS receiver. This rate is not a limitation of the RISS/GPS algorithm or the real-time system implementation. To improve the bounded rate of the navigation solution which is bounded by the 1 Hz rate measurements generated from the robot's wheel encoders and the vehicle's odometer OBDII, high resolution wheel encoders can be used for both the robot and the land vehicle. High resolution encoders will enable the reading of the encoder data at a higher rate while still preserving the accuracy of the measurements which can generate higher rate measurements; lower resolution encoders will cause quantization errors if read at higher rate, furthermore the OBD-II does not provide high rates.

5. The system is required to communicate with a PC via the UART interface to upload the measurements and the computed navigation solution to retrieve, visualize, and analyze the data.

The two approaches that are adopted to test the implemented real-time system are:

1. Using the measurements that are uploaded to the host machine, the solution of the navigation system in real-time is compared to the offline solution computed using the main algorithm that is coded in C which makes use of the uploaded measurements. If the results from both solutions are comparable, then the developed system has succeeded in delivering a real-time navigation solution.

2. The reference solution, which is generated from the NovAtel SPAN unit that contains a GPS receiver and a high-end high-cost tactical grade Honeywell IMU, the HG1700, will be used as a general reference, for assessing the performance of the low-cost developed navigation solution and the offline C solution.

Design Phase

In the design phase, two questions are answered. The first is concerned with the platform to be chosen that will fulfill the specifications of the embedded systems acquired in the first phase. The designer has the option to choose a µC, a DSP, an FPGA, or a combination of these platforms for the application to be developed. Once the platform is chosen, the designer has to decide on the tools that will assist the developers in programming the chosen platform. A processor-based system based on off-the-shelf (OTS) microprocessor solutions is the traditional approach that designers have used in creating embedded systems. These types of traditional processor-based systems are available with a wide range of features and peripherals from a multitude of vendors. The process of selecting an OTS processor that can meet specific cost and functional requirements is time-consuming. It needs all the requirements to be gathered and solidified before the project development begins, which is seldom the case. Also, some of the low-end OTS solutions lack a dedicated floating-point unit which can boost the system performance for math centric algorithms. Due to the mentioned reasons, another approach is sought, which can allow the designer to tailor a processor and a specific set of features and peripherals for the application to be implemented [28].

Based on the above reasons, Xilinx's MicroBlaze [29] soft-core processor was selected to develop the mobile multi-sensor navigation system. Similar to the OTS processors, an FPGA-based soft-core processor like the MicroBlaze can be programmed using high-level languages such as C/C++. MicroBlaze also provides many advantages compared to OTS processor solutions such as [28,30].

1. Customization: The designer of the FPGA-based embedded processor system has the total flexibility to add any combination of peripherals and controllers. A unique set of peripherals can also be designed for specific applications, and the designer has the privilege to add as many peripherals to meet the system requirements, which cannot be done when using an OTS processor. Future system features that didn't have reason to exist in the initial release can also be added due to the space that is provided by an FPGA.

2. Multi-Processor System: As more complex embedded systems can benefit from the existence of multiple processors which can execute the tasks in parallel, by using a soft-core processor accompanying tools, the system designer can easily create a multi-processor based SOC. The only restriction to the number of processors that can be added is the availability of FPGA resources.
3. Hardware Acceleration: Hardware and software concurrent development and co-existence on a single chip, is one of the compelling reasons for choosing a soft core processor. System designers need not to worry, if a segment of the algorithm was found to be a bottleneck, as a custom co-processor or a hardware circuit that utilizes the FPGA parallelism can be designed to eliminate such problem.
4. Obsolescence Mitigation: Supporting extended project life expectance and component obsolescence mitigation is a difficult issue. FPGA soft-processors are an excellent solution in this case since the HDL code for the soft-processor can be purchased, and ownership of the HDL code can fulfill the requirements for product lifespan extension.

The soft-core processor is not without disadvantages. The processor will not operate at the same speeds or have the same performance as a hard-core processor. Unlike an off-the-shelf processor, the hardware platform for the FPGA embedded processor must be designed, where the embedded designer becomes the hardware processor system designer when an FPGA solution is selected. Also, due to the integration of the hardware and software platform design, the design tools are more complex, and the learning curve is steeper. The increased tool complexity and design methodology requires more attention from the embedded designer [30].

The MicroBlaze is a 32-bit reduced instruction set computer (RISC) for use in FPGA designs targeting supported Xilinx Spartan or Virtex families of physical FPGA devices.Figure 5 shows a functional block diagram of the MicroBlaze core [29]. The MicroBlaze soft core processor is highly configurable, allowing the selection of a specific set of features for the system to be designed. Xilinx XPS software tool version 10.1.3 is used which corresponds to MicroBlaze version 7.10.d (For every Xilinx EDK version, there is only one corresponding MicroBlaze version which can be used. Newer MicroBlaze versions can be used only with a new Xilinx EDK release). The fixed feature set of the processor includes: (a) Thirty-two 32-bit general purpose registers (b) 32-bit instruction word with three operands and two addressing modes (c) 32-bit address bus and (d) Single issue pipeline. In addition to these fixed features, the MicroBlaze processor can be parameterized to allow selective enabling of additional functionality [29]. Some of the optional features are introduced in Table 1.

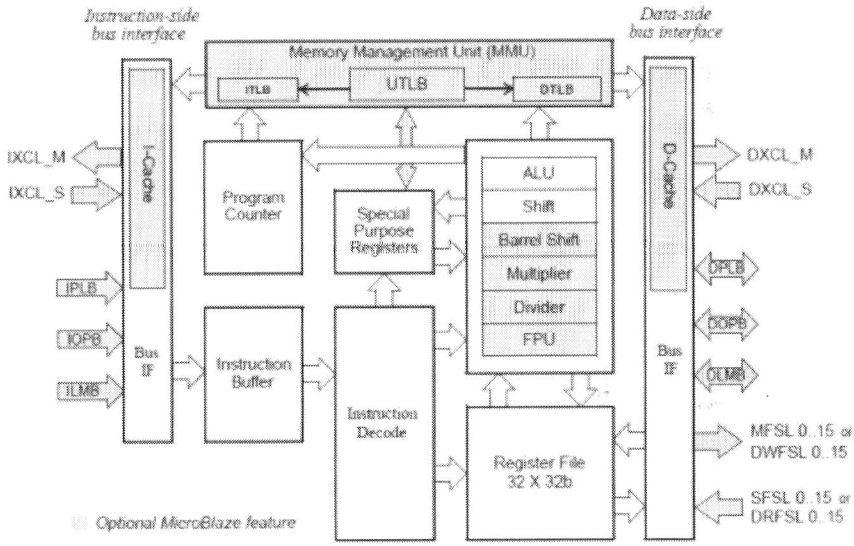

Figure 5. MicroBlaze Core Block Diagram [29].

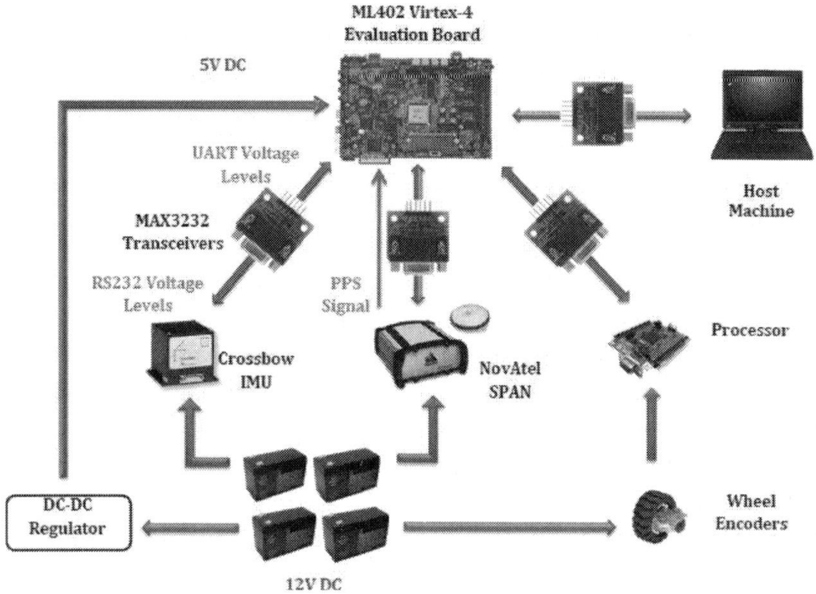

Figure 10. Mobile Robot Equipment Wiring Connections.

A total of fifty trajectories were acquired using the mobile robot around the Royal Military College of Canada over 45 days, which were used in debugging the firmware on the navigation system. This number of trajectories would have been hard to acquire for testing and debugging the system using a land vehicle, which is more costly and requires more man power. Extensive trajectories were performed for both platforms, the mobile robot and the vehicle, and due to the results being consistent and similar, sample trajectories were selected that convey different information and trajectory challenges. 4.2. Mobile Robot—Trajectory (1)

Trajectory (1) was acquired in an open sky area, where no natural GPS outages can be experienced and simulated GPS outages were not inserted. Figure 11 shows the trajectory that was computed by the developed real-time navigation system, and its comparison with the NovAtel SPAN reference solution and the GPS navigation solution. Figure 12 shows the robot's dynamics represented by the azimuth and the horizontal speed from the NovAtel reference over the whole trajectory. The duration of trajectory (1) is 9.6 minutes.

Figure 11. Mobile Robot Trajectory (1): Open Sky Area, with no Natural or Simulated GPS Outages.

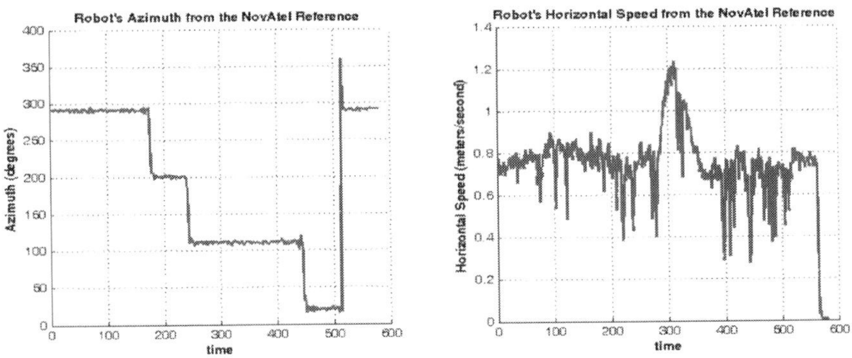

Figure 12. Mobile Robot Dynamics (Azimuth and Horizontal Speed) over Trajectory (1) from the Reference.

In Figure 13, the difference between the positioning solutions from the C double-precision offline algorithm (i.e., The offline solution was computed using the main algorithm that is coded in C on the host machine which makes use of the uploaded measurements from the embedded system via the UART channel) and the C single-precision offline algorithm is presented, where the difference is in the range of 0.5 meter. The results show that the choice of emulating the double-precision on the MicroBlaze core is far more convenient in terms of accuracy, compared to using single-precision arithmetic which would have consumed less memory and less processing time.

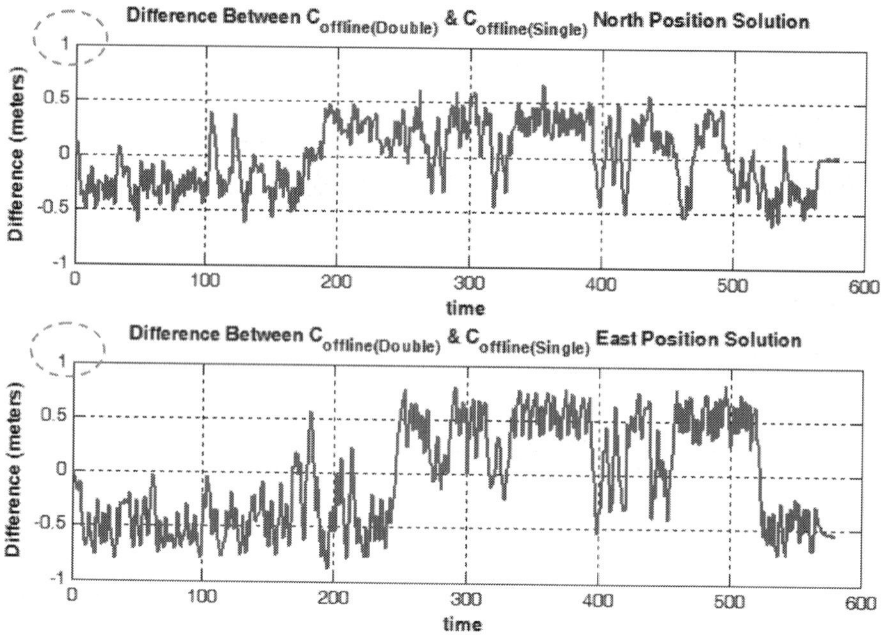

Figure 13. Mobile Robot Trajectory (1): Difference between C Double-Precision Offline and C Single-Precision Offline Solutions.

In Figure 14, the difference between the C double-precision offline solution evaluated on a PC and the MicroBlaze real-time solution with emulated double precision is presented, were the difference is in the magnitude of 10^{-9} meters. The difference can be due to the emulation of double-precision floating-point operations on the single-precision floating point unit present in the MicroBlaze core; however, the difference is negligible and it consequently shows that the real-time system succeeded in providing the same navigation solution as the one generated from the offline system.

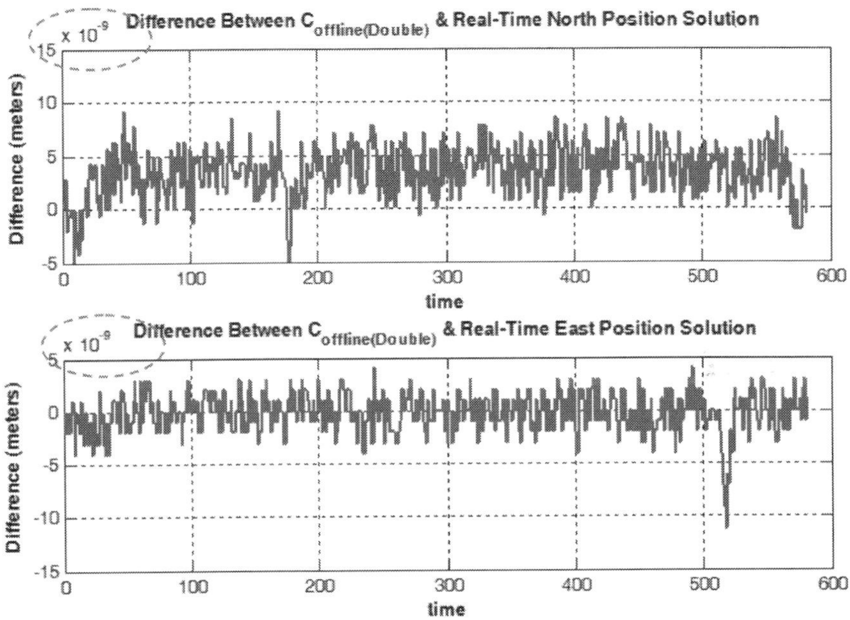

Figure 14. Mobile Robot Trajectory (1): Difference between C Double-Precision Offline and MicroBlaze Real-Time Solutions.

Finally, the difference between the NovAtel SPAN reference solution and the real-time solution is shown in Figure 15. Both solutions are real-time, and the difference is due to the use of the low-cost MEMS-based IMU in the developed solution while the SPAN unit uses the high-end tactical grade HG1700 IMU. In general, the trajectory is open sky trajectory; however the part of the trajectory around 300 seconds from the start is when passing by a bleacher in the "Parade Square" area at the Royal Military College of Canada where some events take place. The GPS positioning solution was not the best at this point and it had about 3 to 4 meters of errors because of multipath, but still the standard deviation provided by the GPS receiver was good.

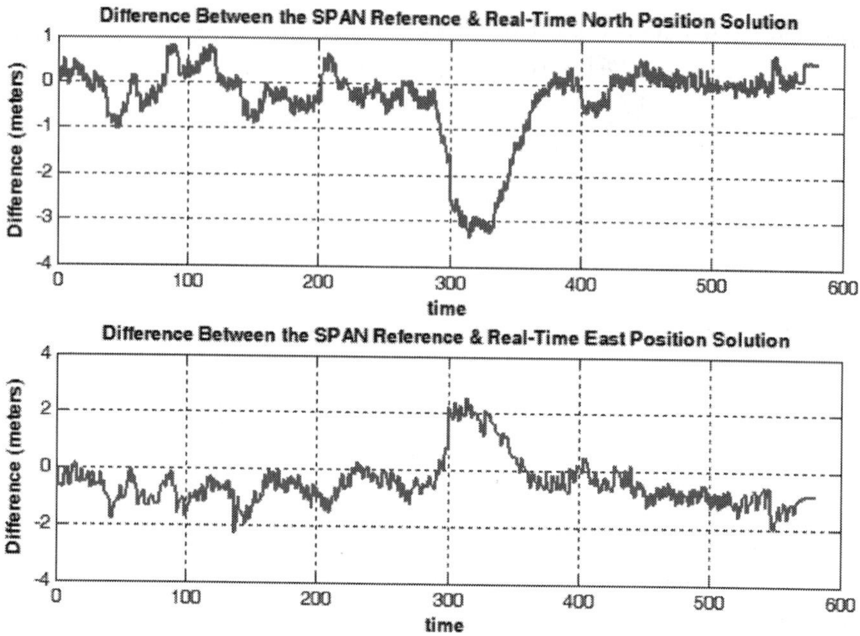

Figure 15. Mobile Robot Trajectory (1): Difference between NovAtel SPAN Reference and MicroBlaze Real Time Solutions.

The off-the-shelf reference solution using a high-end tactical grade IMU, the HG1700, was able to provide the correct positioning solution despite the GPS with undetected multipath, because when using such high-end IMUs the balance of effect (i.e., the balance of the covariance matrices) between the inertial and the GPS solutions inside the integrated solution is different than when using MEMS-based sensors. When using MEMS-based sensors, the integrated solution uses higher values in the covariance matrices because of the large errors present in these sensors which lead to relying more on GPS in the integrated when its standard deviation is good (i.e., its covariance matrix will be good). So, the presented low-cost solution was slightly misled by GPS because of its good standard deviation. This fact can be seen when taking a closer look at Figure 11, in the middle of the bottom side of the rectangular trajectory, where the presented low-cost solution (blue) follows GPS (green) while the high-end reference solution (red) does not. It is to be noted that this scenario is still not a big problem in the presented low-cost solution because other scenarios with noticeably bad GPS position with misleading good standard deviation will not affect the integrated navigation solution because it will be detected and rejected by a GPS assessment routine that uses, in addition to the standard deviation, the motion constraints on land vehicles and speed readings (from wheel encoders or odometer) to detect and reject degraded GPS readings. However, since the GPS error in the current scenario was in the range of 3 to 4 meters and was not conflicting with the motion constraints it was still accepted and gained its way through to update the KF of the integrated solution.

The RMS and maximum errors in the east velocity, north velocity, North position and East position over trajectory (1) entirely are presented in Tables 3 and 4, respectively.

Table 3. RMS Error in the East Velocity, North Velocity, North and East Positions over Mobile Robot Trajectory (1).

East Velocity (m/s)	North Velocity (m/s)	North Position (m)	East Position (m)
RMS Error between the C Double-Precision and C Single-Precision Offline Solutions			
4.12788×10^{-2}	7.98887×10^{-3}	2.96969×10^{-1}	4.91848×10^{-1}
RMS Error between the C Double-Precision Offline and MicroBlaze Real-Time Solutions			
7.83599×10^{-10}	1.44185×10^{-9}	4.16753×10^{-9}	1.85211×10^{-9}
RMS Error between the NovAtel SPAN Reference and MicroBlaze Real-Time Solutions			
9.60578×10^{-2}	1.11987×10^{-1}	9.05270×10^{-1}	9.39925×10^{-1}

Table 4. Maximum Error in the East Velocity, North Velocity, North and East Positions over Mobile Robot Trajectory (1).

East Velocity (m/s)	North Velocity (m/s)	North Position (m)	East Position (m)
Maximum Error between the C Double-Precision and C Single-Precision Offline Solutions			
7.49309×10^{-2}	3.54033×10^{-2}	6.65903×10^{-1}	8.85448×10^{-1}
Maximum Error between the C Double-Precision Offline and MicroBlaze Real-Time Solutions			
1.15944×10^{-8}	2.82925×10^{-8}	9.18873×10^{-9}	1.11810×10^{-8}
Maximum Error between the NovAtel SPAN Reference and MicroBlaze Real-Time Solutions			
4.06407×10^{-1}	4.43351×10^{-1}	3.34354	2.56632

The first row of values in Tables 3 and 4 shows the RMS and maximum errors between the solution generated from both the C double-precision and C single-precision solutions. Again, the values reflect on the fact that the double-precision emulation was important to provide a more accurate navigation solution. This point is further confirmed by the 2nd row of values in Table 3, where it shows that the performance of the offline and real-time systems is essentially the same as the RMS error between solutions from both systems is in the order of 10^{-9} which is negligible. The 3rd row of values in Table 3 shows the RMS error from both the reference and real-time systems where the RMS error in the North and East positions is less than 1 meter.

Mobile Robot—Trajectory (2)

Trajectory (2) is different from trajectory (1) as simulated outages were pre-programmed and inserted in the code at specific instants during the real-time processing. Simulated outages are used to imitate the effect of a GPS outage in areas where natural GPS outages aren't experienced. Outages are introduced in the code to test how the navigation algorithm will perform in the absence of an update from the GPS, i.e., in prediction mode based on the RISS only. The duration of all the simulated outages is 60 seconds. Trajectory (2) follows the same path as trajectory (1), and its duration is 9.3 minutes. Four outages were inserted at the seconds 120, 230, 400 and 525 respectively. Figure 16 shows trajectory (2) on an aerial map, from which the open sky area where the trajectory was acquired can be seen; the simulated outages locations are circled on the map.

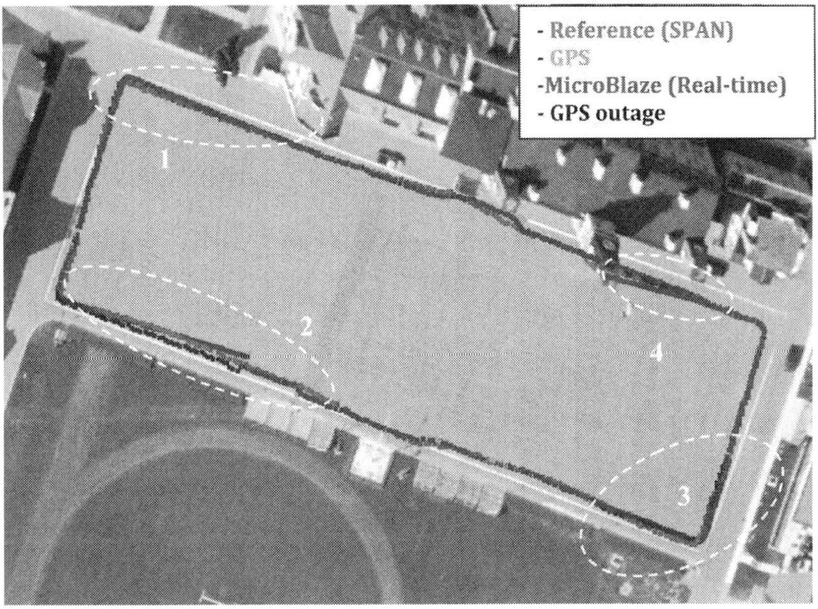

Figure 16. Mobile Robot Trajectory (2): Open Sky Area with Four Simulated GPS Outages.

Trajectory (2) was intended to be the same as trajectory (1) in order to show the effects of the intentionally simulated outages and the performance of the real-time system with respect to trajectory(1) which was analyzed before. Unlike trajectory (1), where the errors where analyzed over the whole trajectory; in trajectory (2), the performance of the real-time system was analyzed within the intentionally inserted outages. In Table 5, the RMS and the maximum horizontal position errors (i.e., The horizontal position error is the square root of the square of the North and East position errors) during the four 60 second outages are presented, which indicate the performance of the real-time solution with respect to the NovAtel SPAN reference solution during the absence of the GPS update. The RMS error between the real-time and reference solutions reaches a maximum of 2.16 meters through outage (4). The performance of the position solution generated from the offline C double and single precisions is again analyzed for the four 60-second outages in trajectory (2) and is shown in Table 6. Again, the difference between both solutions is within a couple of meters indicating the importance of choosing the double-precision arithmetic for the RISS/GPS navigation algorithm instead of the single-precision. The difference between the position solution generated from the offline and the real-time systems is also analyzed within the four outages as shown in Table 6, where the RMS error is in the order of 10^{-6} for outage (2). The difference between the reference and the real-time systems' solutions is then presented where it can be seen that the RMS error in the East and North positions over outage (4) is less than 2 meters to indicate that the performance of the real-time system which contains low-cost MEMS-based inertial sensors is very competitive with respect to the high-end, high-cost reference system.

Table 5. RMS and Maximum Horizontal Position Error for the Four 60 seconds Outages Introduced in Mobile Robot Trajectory (2).

Outage (1)	Outage (2)	Outage (3)	Outage (4)
RMS Horizontal Position Error between the SPAN Reference and MicroBlaze Real-Time Solutions (m)			
9.38501×10^{-1}	1.40681	1.06658	2.16077
Maximum Horizontal Position Error between the SPAN Reference and MicroBlaze Real-Time Solutions (m)			
1.65584	2.48860	1.79489	3.08822

Table 6. RMS Error in the Latitude and Longitude Solutions over the Four Outages introduced in Mobile Robot Trajectory (2).

	North position (m)	East position (m)
Outage (1)	RMS Error between the C Double-Precision and C Single-Precision Offline Solutions	
	3.00674	6.42866
	RMS Error between the C Double-Precision Offline and MicroBlaze Real-Time Solutions	
	1.37906×10^{-8}	4.92348×10^{-8}
	RMS Error between the NovAtel SPAN Reference and MicroBlaze Real-Time Solutions	

	7.91314×10^{-1}	**5.04586×10^{-1}**
Outage (2)	RMS Error between the C Double-Precision and C Single-Precision Offline Solutions	
	3.39406	**4.07745**
	RMS Error between the C Double-Precision Offline and MicroBlaze Real-Time Solutions	
	1.38649×10^{-6}	**4.47223×10^{-7}**
	RMS Error between the NovAtel SPAN Reference and MicroBlaze Real-Time Solutions	
	4.42573×10^{-1}	**1.33538**
Outage (3)	RMS Error between the C Double-Precision and C Single-Precision Offline Solutions	
	1.45689	**3.67700**
	RMS Error between the C Double-Precision Offline and MicroBlaze Real-Time Solutions	
	3.14829×10^{-7}	**2.40989×10^{-7}**
	RMS Error between the NovAtel SPAN Reference and MicroBlaze Real-Time Solutions	
	7.32207×10^{-1}	**7.75539×10^{-1}**
Outage (4)	RMS Error between the C Double-Precision and C Single-Precision Offline Solutions	
	8.03891×10^{-1}	**3.30091**
	RMS Error between the C Double-Precision Offline and MicroBlaze Real-Time Solutions	
	6.60678×10^{-9}	**2.16253×10^{-9}**
	RMS Error between the NovAtel SPAN Reference and MicroBlaze Real-Time Solutions	
	1.84932	**1.11755**

Mobile Robot—Trajectory (3)

Figure 17 shows trajectory (3), which is different from the previously presented trajectories, in having a portion of the trajectory acquired inside of a building. A natural GPS outage of 100 seconds occurred, which is equivalent to the time taken by the mobile robot from the entrance to the exit of the building. The natural outage was detected by the system which acted accordingly in providing a navigation solution in prediction only mode based on RISS readings; the solution still shows a very competitive performance with respect to the high-end, high-cost NovAtel SPAN reference system. The RMS error in the horizontal position during the outage is 2.39 meters, while the maximum error is 10.25 meters. Figures 18 and 19 show the error in North and East positions between the high-end reference solution and the developed low-cost real-time solution.

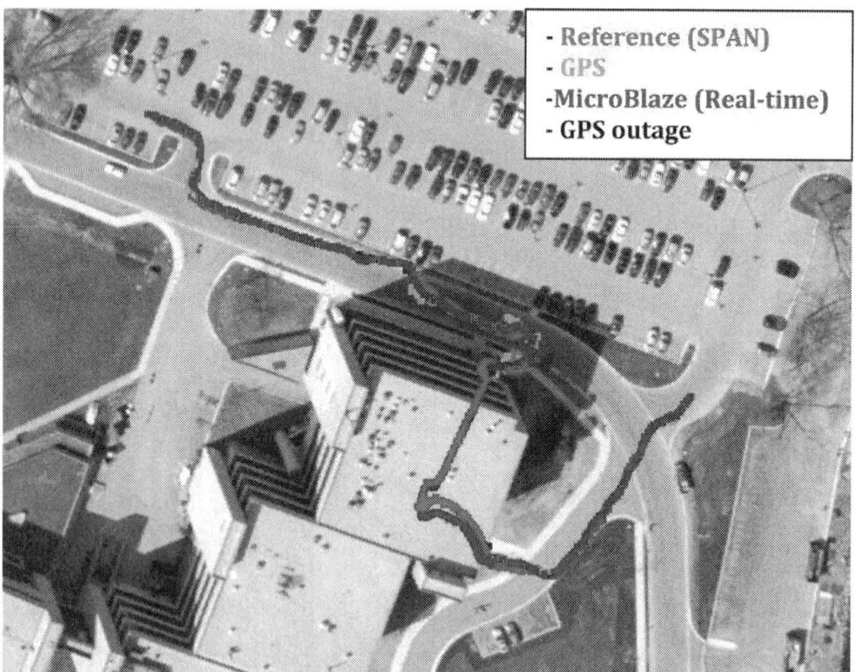

Figure 17. Mobile Robot Trajectory (3): Inside a building with a natural GPS Outage of 100 seconds duration.

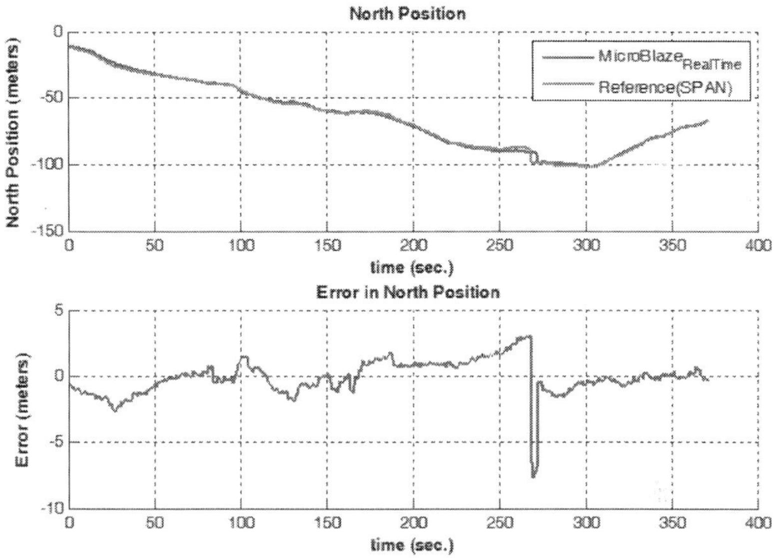

Figure 18. North Position Solution and Error in North Position between the Real-time and Reference Solutions during Trajectory (3).

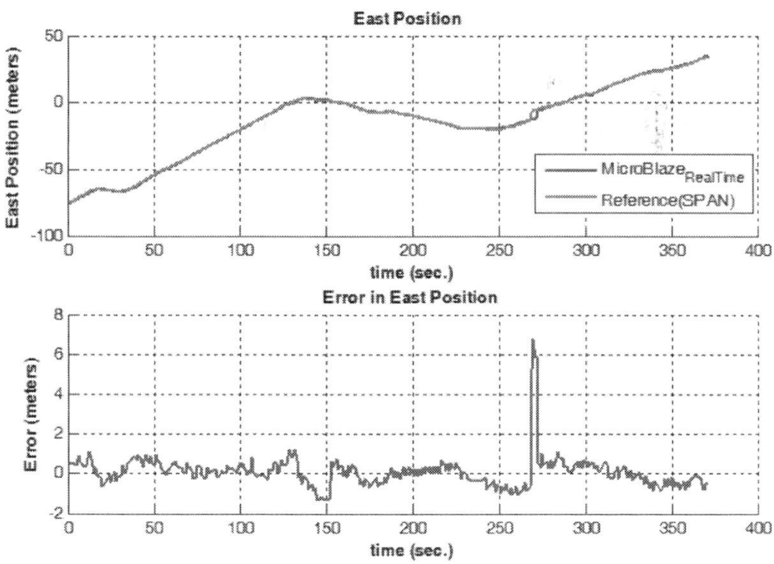

Figure 19. East Position Solution and Error in East Position between the Real-time and Reference Solutions during Trajectory (3).

Land Vehicle Testing Platform

After the firmware of the navigation system was debugged on the mobile robot and proven to give the same results as the C-code offline solution, the next step was to test the navigation system on a land vehicle where the wheel encoders on the mobile robot are replaced with the ElmScan vehicle speed scan tool that reads the vehicle speed readings through OBD-II interface. The experimental setup of the sensors and the FPGA evaluation kit on the land vehicle platform is shown in Figure 20.

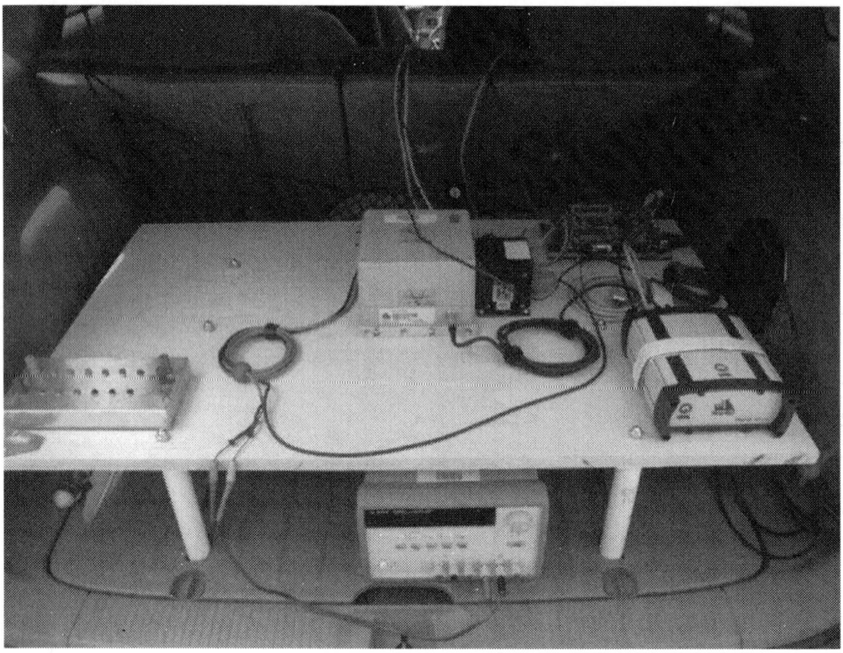

Figure 20. The Land Vehicle Experimental Setup.

Figure 21 shows the wiring connections of the sensors with the ML402 evaluation board featuring Xilinx Virtex-4 FPGA on which the MicroBlaze soft-core processor is running. The power connections to the sensors and the evaluation board are also shown.

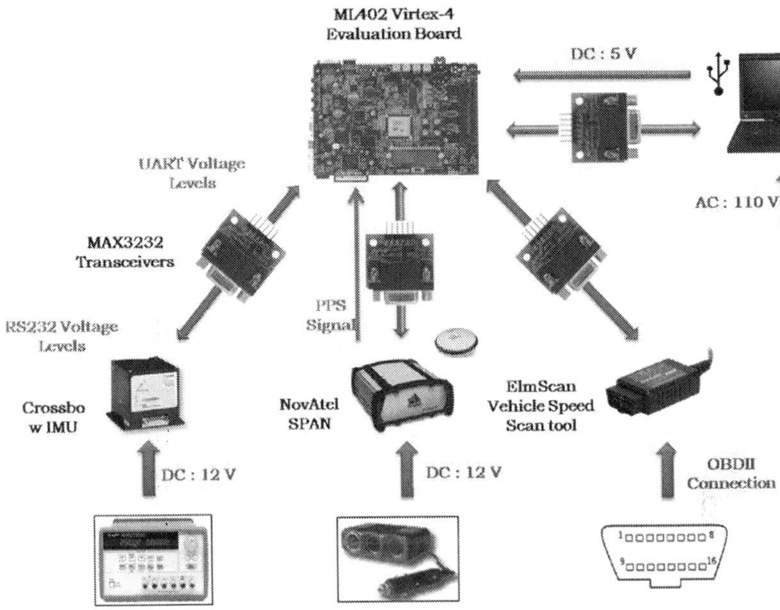

Figure 21. Land Vehicle Equipment Wiring Connections.

Land Vehicle—Trajectory (1)

Trajectory (1) was acquired in an open sky area in Kingston around Queen's University, where a small number of natural GPS outages can be experienced and simulated GPS outages were not inserted. Figure 22 shows the trajectory that was computed by the developed real-time navigation system, and its comparison with the NovAtel SPAN reference solution and the GPS navigation solution. Figure 23 shows the vehicle's dynamics over the whole trajectory from the NovAtel reference solution. The duration of the trajectory is 9.2 minutes.

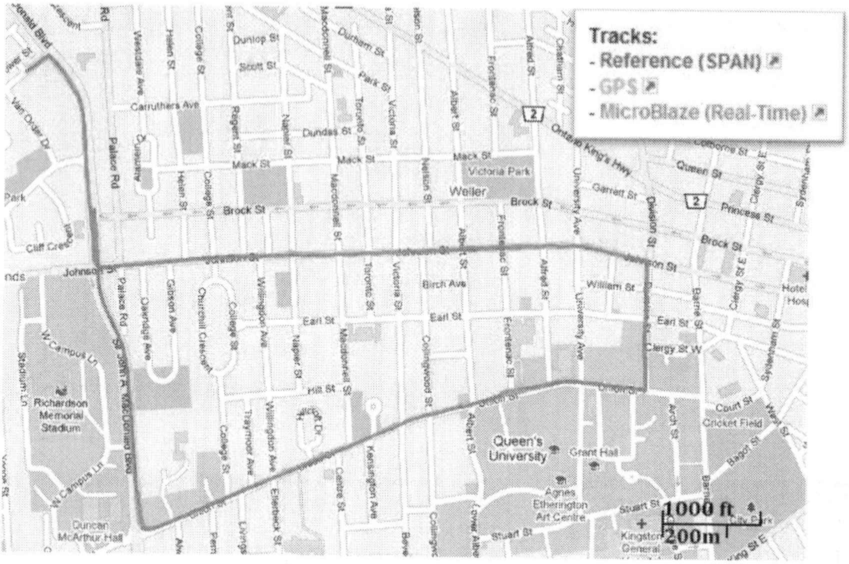

Figure 22. Land Vehicle Trajectory (1): Open Sky Area, with Minimum Natural Outages and no Simulated GPS Outages.

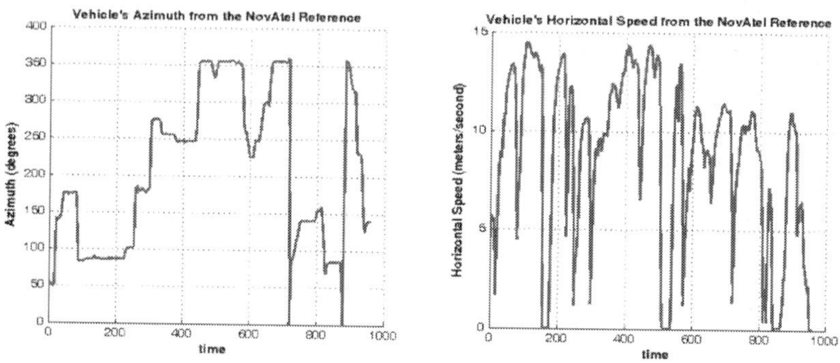

Figure 23. Land Vehicle Dynamics (Azimuth and Horizontal Speed) for Trajectory (1).

The differences between the offline, real-time and reference navigation solutions are presented in Tables 7 and 8 where the RMS and maximum errors in the East velocity, North velocity, East position, and North position over the whole trajectory are shown. The results for the land vehicle are comparable to the results for the mobile robot, from which it can be shown that the performance of the offline and real-time systems is very similar as the RMS error between the solutions generated from both systems is in the order of 10^{-7}. It is to be noted that although the trajectory was taken in nearly open-sky, as mentioned previously a small number of natural GPS outages was experienced and GPS was also rejected in some areas mainly due to multipath caused by the trees on both sides of the street in these areas. This is the reason for the position error seen in Tables 7 and 8.

Table 7. RMS Error in the East Velocity, North Velocity, North and East Positions over Land Vehicle Trajectory (1).

East Velocity (m/s)	North Velocity (m/s)	North Position (m)	East Position (m)
RMS Error between the C Double-Precision and C Single-Precision Offline Solutions			
1.78523×10^{-1}	9.60298×10^{-3}	3.76213×10^{-1}	4.98008×10^{-1}
RMS Error between the C Double-Precision Offline and MicroBlaze Real-Time Solutions			
5.85936×10^{-8}	1.48418×10^{-8}	1.19570×10^{-7}	3.67245×10^{-7}
RMS Error between the NovAtel SPAN Reference and MicroBlaze Real-Time Solutions			
6.73738×10^{-1}	5.757354×10^{-1}	6.42297	7.79358

Table 8. Maximum Error in the East Velocity, North Velocity, North and East Positions over Land Vehicle Trajectory (1).

East Velocity (m/s)	North Velocity (m/s)	North Position (m)	East Position (m)
Maximum Error between the C Double-Precision and C Single-Precision Offline Solutions			
1.405039×10^{-1}	6.68926×10^{-2}	1.24497	2.82222
Maximum Error between the C Double-Precision Offline and MicroBlaze Real-Time Solutions			
4.55655×10^{-7}	1.47957×10^{-7}	1.02348×10^{-6}	3.84857×10^{-6}
Maximum Error between the NovAtel SPAN Reference and MicroBlaze Real-Time Solutions			
4.50865	3.02078	2.01342×10	3.63886×10

Table 9. RMS and Maximum Horizontal Position Error for the Four 60 seconds Outages Introduced in Land Vehicle Trajectory (2).

Outage (1)	Outage (2)	Outage (3)	Outage (4)
RMS Horizontal Position Error between the SPAN Reference and MicroBlaze Real-Time Solutions (m)			
8.91707	1.40813 × 10	1.12782 × 10	1.63253 × 10
Maximum Horizontal Position Error between the SPAN Reference and MicroBlaze Real-Time Solutions (m)			
1.66452 × 10	2.37866 × 10	1.57619 × 10	2.39850 × 10

CONCLUSIONS

This paper discussed the process and challenges of realizing a mobile multi-sensor navigation algorithm such as the KF for 2D RISS/GPS integration algorithm. An embedded system design model was chosen to act as a framework for the work flow to be carried through the system life cycle starting from the system specification phase and ending with the system release. The realized system is capable of interfacing and communicating with a GPS receiver, a gyroscope and a vehicle's odometer or a robot's wheel encoders, synchronizing the sensors' measurements with respect to the PPS signal and then applies the navigation algorithm yielding a reliable and accurate integrated navigation solution. Xilinx's soft-core processor, MicroBlaze, on Virtex-4 FPGA was selected as the most suitable candidate for implementing the navigation system, where it provides the flexibility to choose or implement a set of features and peripherals that are tailored to the navigation system. The MicroBlaze also provides a single-precision floating point unit which was used to emulate the double-precision arithmetic embodied in the navigation algorithm, as the accuracy of the double-precision solution was higher than that of the single-precision solution. The error between the real-time emulated double-precision solutions when compared to the offline double-precision solution was in the range of 10^{-9} meters. The navigation system on the high-density Virtex-4 FPGA, utilized 20% of the total available logic cells 33.3% of the total available BRAM, which means that a lower-density FPGA can be used for implementing the system. The developed navigation system was tested first on a mobile robot to reveal system bugs and integration problems, and then on a land vehicle testing platform for further testing. The real-time solution from the implemented system when compared to the solution of a high-end navigation system, proved to be successful in providing a competitive consistent real-time navigation solution.

Using FPGA-based processors offer system designers the maximum flexibility to customize the processor to specific applications. The designers have even more flexibility to use multi-MicroBlaze cores for parallel computation. Co-designing software and hardware is also another option offered where designers can implement the bottleneck segments of the algorithm as a custom coprocessor or a hardware circuit that speeds up the algorithm. All these options promote the utilization of soft-core processors in a more attractive and feasible approach than the traditional OTS approach.

This paper doesn't promote a soft-core processor from a specific vendor; however, it promotes using soft-core processors for implementing multi-sensor navigation systems as the future platform due to the offered advantages discussed in the paper. The paper demonstrated in length how embedded system design and development for FPGA-based processors is different than the traditional OTS processors, with an emphasis on navigation applications.

This research was supported in part by research grants from Natural Sciences and Engineering Research Council (NSERC), Geomatics for Informed Decision (GEOIDE) Network Centers of Excellence, and Defence Research and Development Canada (DRDC) Ottawa. The equipment was acquired by research funds from Canada Foundation for Innovation, Ontario Innovation Trust and the Royal Military College of Canada.

REFERENCES

1. Bekir, E. Introduction to Modern Navigation Systems; World Scientific Publishing Company: Singapore, 2007.
2. El-Rabbany, A. Introduction to GPS: The Global Positioning System; Artech House Publishers: Boston, MA, USA, 2002.

3. Woodman, O.J. An Introduction to Inertial Navigation. Available online: http://www.cl.cam.ac.uk/~ojw28/ins.html (accessed on 27 September 2011)..
4. Maluf, N.; Williams, K. An Introduction to Microelectromechanical Systems Engineering, 2nd ed; Artech House Publishers: Boston, MA, USA, 2004.
5. Farrell, J. Aided Navigation: GPS with High Rate Sensors;; McGraw-Hill Professional: New York, NY, USA, 2008.
6. Iqbal, U.; Okou, A.F.; Noureldin, A. An integrated reduced inertial sensor system—RISS/GPS for land vehicle. Proceedings of the Position, Location and Navigation Symposium (2008 IEEE/ION), Monterey, CA, USA, 5–8 May 2008; pp. 1014–1021.
7. Iqbal, U.; Noureldin, A. Integrated Reduced Inertial Sensor System/GPS for Vehicle Navigation: Multi-Sensor Positioning System for Land Applications Involving Single-Axis Gyroscope ... Vehicle Odometer and Integrated with GPS; VDM Verlag Dr. Müller: Saarbrücken, Germany, 2010.
8. Bevly, D.M. GNSS for Vehicle Control; Artech House Publishers: Boston, MA, USA, 2009; pp. 1-266.
9. Carlson, R.; Gerdes, J.; Powell, J. Error Sources When land vehicle dead reckoning with differential wheelspeeds. Navigation **2004**, 51, 13–27.
10. Dissanayake, G.; Sukkarieh, S.; Nebot, E.; Durrant-Whyte, H. The aiding of a low-cost strapdown inertial measurement unit using vehicle model constraints for land vehicle applications. IEEE Trans. Robot. Autom **2001**, 17, 731–747.
11. Obradovic, D.; Lenz, H.; Schupfner, M. Fusion of sensor data in siemens car navigation system. IEEE Trans. Veh. Technol **2007**, 56, 43–50.
12. Skog, I.; Handel, P. In-car positioning and navigation—A survey. IEEE Trans. Intell. Transp. Syst **2009**, 10, 4–21.
13. Marwedel, P. Embedded System Design; Springer: Berlin, Germany, 2005.
14. Noergaard, T. Embedded Systems Architecture: A Comprehensive Guide for Engineers and Programmers; Newnes: Waltham, MA, USA, 2005.
15. Wolf, W. High-Performance Embedded Computing: Architectures, Applications, and Methodologies; Morgan Kaufmann: Waltham, MA, USA, 2006.
16. Anemaet, P.; As., T.V. Microprocessors Soft-Cores: An Evaluation of Design Methods and Concepts on FPGAs. Available online: http://pretopia.net/files/paper_softcores.pdf (accessed on 27 September 2011)..

17. Dubey, R. Introduction to Embedded System Design Using Field Programmable Gate Arrays; Springer: Berlin, Germany, 2008.
18. Li, Q.; Yao, C. Real-Time Concepts for Embedded Systems; CMP Books: San Francisco, CA, USA, 2003; p. xii.
19. Choosing a DSP Processor; Berkeley Design Technology, Inc.: Berkeley, CA, USA, 2000; Available online: http://www.bdti.com/MyBDTI/pubs/choose_2000.pdf(accessed on 27 September 2011)..
20. Eyre, J.; Bier, J. The Evolution of DSP Processors, 2000. Available online: http://www.bdti.com/articles/info_articles.htm (accessed on 27 September 2011)..
21. Lapsley, P. DSP Processor Fundamentals: Architectures and Features; IEEE Press: New York, NY, USA, 1997; p. xiii.
22. Stringham, G. Hardware/Firmware Interface Design: Best Practices for Improving Embedded Systems Development; Newnes: Waltham, MA, USA, 2009.
23. Grewal, M.S.; Weill, L.R.; Andrews, A.P. Global Positioning Systems, Inertial Navigation, and Integration, 2nd ed; Wiley-Interscience: Hoboken, NJ, USA, 2007; p. xxvi.
24. SPAN Technology System User Manual; NovAtel Inc: Goleta, CA, USA, 2005. Available online: http://www.novatel.com/Documents/Manuals/om-20000062.pdf(accessed on 27 September 2011)..
25. OEM4 Family. Installation and Operation; OM-20000046 Rev 19, User Manual—Volume 1. NovAtel Inc: Goleta, CA, USA. Available online: http://www.novatel.com/support/firmware-software-and-manuals/product-manuals-and-doc-updates/archive-oem4-oem3-oem2/ (accessed on 27 September 2011)..
26. IMU300—6DOF Inertial Measurement Unit; Crossbow Technology Inc.: San Jose, CA, USA. Available online: http://www.instrumentation.it/main/pdf/crossbow/IMU300CC_ID.pdf (accessed on 27 September 2011)..
27. ELM327 OBD to RS232 Interpreter. Available online: www.elmelectronics.com/DSheets/ELM327DS.pdf (accessed on 27 September 2011)..
28. Arbinger, D.; Erdmann, J. Designing with an embedded soft-core processor. Available online: http:// www.eetimes.com/design/signal-processing-dsp/4006632/Designing-with-an-embedded-soft-core-processor (accessed on 27 September 2011)..
29. Xilinx Inc. MicroBlaze Processor Reference Guide Embedded Development Kit EDK 10.1. Available online: http://www.xilinx.com/ise/embedded/edk_docs.htm(accessed on 27 September 2011)..

30. Fletcher, B.H. FPGA Embedded Processors—Revealing True System Performance. Proceedings of the Embedded Systems Conference, San Francisco, CA, USA, 6–10 March 2005.
31. Xilinx Inc. EDK Concepts, Tools, and Techniques A Hands-on Guide to Effective Embedded System Design. Available online: http://www.xilinx.com/ise/embedded/edk_docs.htm (ac cessed on 27 September 2011)..
32. Ding, W.; Wang, J.; Li, Y.; Mumford, P.; Rizos, C. Time synchronization error and calibration in integrated GPS/INS systems. ETRI J **2008**, 30, 59–67.

CITATION

Walid Farid Abdelfatah , Jacques Georgy , Umar Iqbal and Aboelmagd Noureldin, FPGA-Based Real-Time Embedded System for RISS/GPS Integrated Navigation, doi:10.3390/s120100115.

CHAPTER 8

Microstructured Optical Fiber Sensors Embedded in a Laminate Composite for Smart Material Applications

Camille Sonnenfeld [1,2,], Sanne Sulejmani [1], Thomas Geernaert [1], Sophie Eve [2], Nicolas Lammens [3], Geert Luyckx [3], Eli Voet [3], Joris Degrieck [3], Waclaw Urbanczyk [4], Pawel Mergo [5], Martin Becker [6], Hartmut Bartelt [6], Francis Berghmans [1] and Hugo Thienpont [1]*

[1]Vrije Universiteit Brussel, Pleinlaan 2, B-1050 Brussel, Belgium; E-Mails: ssulejma@vub.ac.be(S.S.); tgeernae@vub.ac.be (T.G.); fberghma@vub.ac.be (F.B.); hthienpo@vub.ac.be (H.T.)

[2]CRISMAT CNRT-Matériaux/ENSICAEN, UMR 6508, UMS 3318, 6, Boulevard du Maréchal Juin, 14050 Caen Cedex 04, France; E-Mail: sophie.eve@ensicaen.fr

[3]Universiteit Gent, Sint-Pietersnieuwstraat 41, B-9000 Gent, Belgium; E-Mails:Nicolas.Lammens@ugent.be (N.L); Geert.luyckx@ugent.be (G.L); evoet@fos-s.com (E.V.);Joris.Degrieck@UGent.be (J.D.);

[4]Wroclaw University of Technology, Wybrzeze Wyspianskiego 27, 50-370 Wroclaw, Poland; E-Mail:Waclaw.Urbanczyk@pwr.wroc.pl

[5]Marie Curie-Sklodowska University, Pl. M. Curie-Sklodowskiej 3, 20-031 Lublin, Poland; E-Mail:pawel.mergo@poczta.umcs.lublin.pl

[6]Institute of Photonic Technology, Albert-Einstein-Straße 9, D-07745 Jena, Germany; E-Mails:martin.becker@ipht-jena.de (M.B.); hartmut.bartelt@ipht-jena.de (H.B.)

ABSTRACT

Fiber Bragg gratings written in highly birefringent microstructured optical fiber with a dedicated design are embedded in a composite fiber-reinforced polymer. The Bragg peak wavelength shifts are measured under controlled axial and transversal strain and during thermal cycling of the composite sample. We obtain a sensitivity to transversal strain that exceeds values reported earlier in literature by one order of magnitude. Our results evidence the relevance of using microstructured optical fibers for structural integrity monitoring of composite material structures.

INTRODUCTION

Monitoring the internal strain state of fiber-reinforced polymer materials has become an important issue since in-service strain monitoring of civil engineering and aeronautic structures can lead to improved safety and better control over costs [1]. Fiber Bragg grating (FBG) based sensors are excellent candidates for that purpose as they can be embedded in different materials for smart structure applications. These sensors combine many advantages over conventional electrical sensor configurations, such as for example their small size, their immunity to electromagnetic interference, their multiplexing capabilities and self-referencing ability together with an often linear response that is encoded in the change of their reflected resonance wavelength.

To monitor the structural health of composite fiber-reinforced polymer (CFRP), one ideally needs a complete mapping of the internal strain field of the material using multi-axial strain sensors. One also needs to distinguish between the strain occurring in the axial and in the transverse directions. Monitoring the strain in the transverse direction of a CFRP in a laminated configuration is indeed an essential issue. In such structures reinforcement fibers enhance the mechanical strength of the composite in the plane of the structure, but they suffer from fragility in the transverse direction. One therefore needs to develop a system that is able to sense the out-of-the plane strains in composite laminates.

The possibility of embedding optical fibers in anisotropic materials has already been demonstrated [2]. Nevertheless most of the work published so far has focused on axial load and temperature variations, but did not investigate in detail the out-of-plane strain which is a key parameter in the structural integrity of anisotropic materials. The sensitivity to transversal strain also remained considerably lower than the sensitivity to axial strain [3,4]. In recent years substantial research efforts also attempted at developing multi-parameter FBG sensors. Several reports described the characterization of gratings subjected to transversal strains [5,6]. The Bragg gratings were written either in polarization maintaining-fibers [4,7,8] or in microstructured optical fiber [9]. However distinguishing between transversal and axial strain and correcting for temperature cross-sensitivity still required to use a second grating [8].

Several studies indeed relied on two FBGs written in the same fiber: the first grating is sensitive to all external perturbations (strain and temperature) while the second grating remains isolated from (part of) the external mechanical strain. This approach allows precisely defining the influence of each perturbation [10,11] but the presence of the capillary used to isolate the second grating from mechanical load disturbs the structure under test.

In our work we demonstrate a new optical fiber sensor specifically designed to identify transverse strain components in a CFRP with a much higher sensitivity than ever achieved before, and with a response that is independent of temperature variations. To do so we use FBGs inscribed in a highly birefringent microstructured optical fiber (MOF). Our paper is structured as follows. In the next section we summarize the theory of FBGs and we characterize the

Bragg grating inscribed in the MOF. In Section 3 we theoretically and mechanically characterize the response of the bare free standing optical fiber to a transverse line load. In the fourth section we describe the embedding procedure and we assess the behavior of the embedded fiber under several loading conditions (temperature, axial and transversal tests). We eventually compare our results with those already reported in literature. Finally we conclude on the possible use of integrated MOFs for precisely monitoring transverse strain fields in laminated composite structures.

DESCRIPTION OF THE MICROSTRUCTURED OPTICAL FIBER AND CHARACTERISTICS OF THE BRAGG GRATING

Fiber Bragg Grating Sensor Theory

A fiber Bragg grating is a periodic modulation of the refractive index of the fiber core, extended over a limited length of the fiber (typically a few mm). A FBG acts as a wavelength selective filter for light traveling along the fiber: it reflects light with a spectrum centered on the Bragg wavelength λ_B, which is given by the following equation:

$$\lambda_B = 2.\ n_{eff}.\ \Lambda \tag{1}$$

where Λ is the grating period and n_{eff} is the mean effective refractive index in the core region [12]. In a birefringent fiber two orthogonally polarized modes can propagate along the fiber with different phase velocities. A Bragg grating in such a fiber will return two reflection peaks (at wavelengths λ_{B1} and λ_{B2}) corresponding to the two orthogonally polarized modes travelling along respectively the slow and fast axis of the optical fiber. The spectral distance $\Delta\lambda_B$ between these two peaks, to which we will refer as the wavelength separation, is proportional to the phase modal birefringence B:

$$\Delta\lambda_B = 2.\Lambda.B \tag{2}$$

$$B = n_1 = n_2 \tag{3}$$

where n_1 and n_2 are the effective indices of the two orthogonally polarized modes. Both the effective indices and the grating period will be affected by temperature and by applied mechanical strain. To determine the resulting Bragg wavelength shift of FBGs written in mechanically anisotropic optical fibers, Kim *et al.* [13] made the assumption that most of the energy of the fundamental mode propagating in the fiber is contained within the core and hence the principal strains at the centre of the fiber are sufficient to determine this wavelength shift. Under this approximation and when assuming constant temperature, the wavelength shifts can be derived from the total strain field present in the center of the fiber core as:

$$\frac{\Delta\lambda_{B,1}}{\lambda_{B,1}} = \varepsilon_3 - \frac{1}{2} n_1^2 \left[p_{11}\varepsilon_1 + p_{12}\left(\varepsilon_2 + \varepsilon_3\right)\right] \tag{4}$$

$$\frac{\Delta\lambda_{B,2}}{\lambda_{B,2}} = \varepsilon_3 - \frac{1}{2} n_2^2 \left[p_{11}\varepsilon_2 + p_{12}\left(\varepsilon_1 + \varepsilon_3\right)\right] \tag{5}$$

where ε_1, ε_2 and ε_3 are the principal strain components along the axes of the fiber (ε_3 refers to the axial direction), p_{11} and p_{12} stand for the strain-optic coefficients [14], $\lambda_{B1,2}$ denote the initial unstrained wavelengths of the Bragg peaks for each polarization mode and

$\Delta\lambda_{B,1,2}$ are the Bragg wavelength peak shifts. By applying mechanical strain to the grating, the phase modal birefringence is modified, which in turn induces a measurable variation of the wavelength separation.

Fiber Microstructure Topology and Description of the Bragg Grating

In this report we rely on a microstructured optical fiber. The cross-section of the silica fiber contains a doped core with a GeO_2 concentration of 7.4 mol% and a distribution of micro air-holes running along the fiber length [Figure 1(a)]. It is already well known that by modifying the air-hole microstructure, i.e. by changing the diameter, or the distance between the air-holes and their location in the fiber cross-section, one can tailor the guiding properties of the fiber for specific applications and enhance its sensitivity to particular physical quantities [15]. The air-hole topology of the MOF under consideration has already demonstrated a high sensitivity to hydrostatic pressure [16]. Here we focus on its high sensitivity to transversal loading. Moreover, the phase modal birefringence in this MOF is inherently insensitive to temperature as reported in [16] which helps avoiding complex temperature compensation systems.

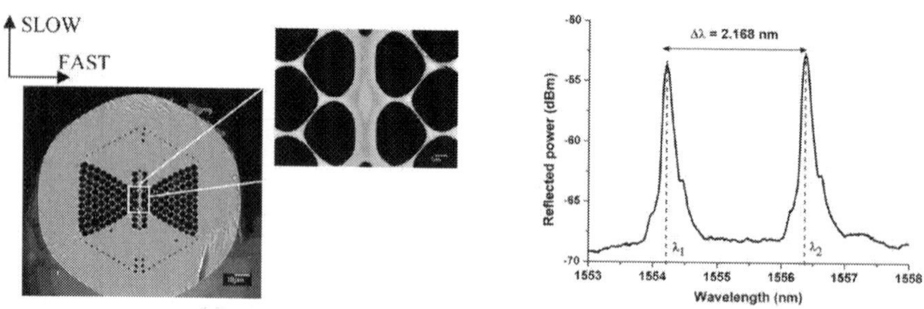

(a) (b)

Figure 1. (a) Scanning electron microscope micrograph of a cross-section of the studied MOF and close up of the core region. **(b)** Bragg grating spectrum inscribed in the fiber core.

The function of the germanium doped fiber core is to allow using conventional ultraviolet fiber Bragg grating inscription methods. The FBGs were written with pulses of UV radiation originating from a Kr:F laser operating at 248 nm and with an interferometric Talbot system [17]. The writing of Bragg gratings in the germanium doped core of MOFs using this inscription technique has already been reported in [18]. The gratings are 5 mm in length with an averaged linewidth of 201 pm. The resulting Bragg spectrum shown in Figure 1(b) exhibits two clearly identified Bragg peaks. The peak separation is about 2 nm at λ = 1,550 nm which corresponds to a modal birefringence of about 2×10^{-3}. This value is almost one order of magnitude larger than the values for conventional birefringent optical fibers, e.g., as reported in [19].

SIMULATIONS AND EXPERIMENTS FOR THE BARE MICROSTRUCTURED OPTICAL FIBER

The sensing potential of a FBG in the proposed MOF can already be demonstrated by characterizing its response to a transversal line load. The experiments were made by compressing the grating between a metal and a glass plate in a dedicated test bench for the calibration of FBG-based sensors [20]. We applied a transversal line load of 0.2 N/mm to the fiber and monitored the wavelength peak variation using an amplified spontaneous emission source, a circulator and a commercially available FBG interrogator with a peak wavelength detection resolution of 1 pm.

Since the air-hole topology is highly asymmetric, the transversal line load sensitivity changes with the angular orientation of the applied force with respect to the fiber cross-section. The grating sensitivity is determined as the slope of a linear fit of the Bragg peak wavelength shift *versus* the applied transversal line load for various angular orientations in steps of 15° [Figure 2(a)]. The typical values of the

R^2coefficients of the linear fits were above 0.995. The fiber response according to the force orientation exhibits, as expected, a periodic behavior [Figure 2(b)]. We can point out the differences in the fiber response at 0° and at 360° while the fiber orientations are theoretically identical. These different values are most likely due to the difficult control of the fiber orientation and hence to its uncertainty during the experiments. The highest sensitivity of about −370 pm/(N/mm) is obtained when the load is applied along the slow axis of the fiber. The large side microstructure is designed to guide the applied load along this eigenaxis to confine the stress into the core, which increases the material birefringence and hence the sensitivity. We can compare our sensitivity to results published in literature. It is twice as large as that obtained using conventional highly birefringent (HB) fibers such as bow-tie fiber [160 pm/(N/mm) [19]]. In 2007 the sensing characteristics of a 220 μm diameter side-hole fiber were investigated [7]. The Bragg peak separation sensitivity reached 176 pm/(N/mm) which is half the value found in our experiments. In 2009 the Bragg peak separation sensitivity of a simple microstructured fiber tested in diametrical compression was reported to be about 100 pm/(N/mm) [20].

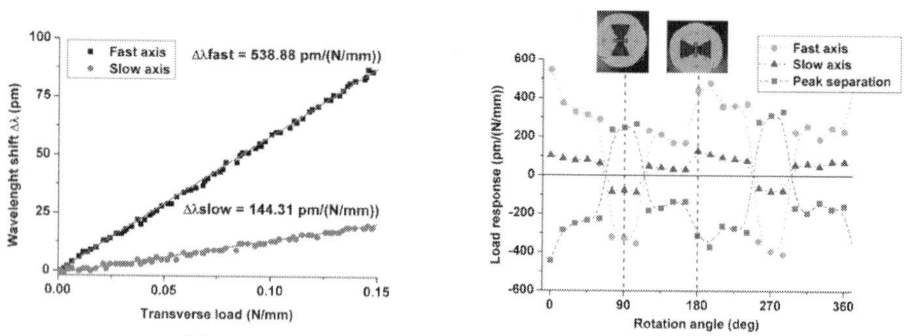

(a) (b)

Figure 2. (a) Wavelength shift of the Bragg peaks *versus*transversal line load for one particular fiber orientation. The red solid lines represent the linear fit. **(b)** Transverse line load sensitivity plotted against angle of rotation for FBG in the fiber. The dashed vertical lines indicate the maximum values.

A numerical simulation of the FBG sensitivity to transversal line load was performed with the commercially available COMSOL Multiphysics software in a two-dimensional finite element model [16,20,21]. The exact dimensions of the simulated fiber have been extracted from a scanning electron microscope image of the fabricated fiber. The simulation combines mechanical calculations using the plane strain approximation and optical calculations where a full-vector perpendicular wave representation is used to solve for the optical modes. The mechanical calculations attempt at determining the stress distribution in the fiber which is subsequently used in the optical calculations to find the effective refractive indices of the fundamental optical modes of the fiber.

The results for the maximum and minimum sensitivity obtained from the finite element model show a reasonable agreement with the experimental results (Figure 3). The deviations are most likely due to the uncertainty on the angular orientation of the optical fiber. As can be seen in Figure 1(a) the outer shape of the cladding is not perfectly circular, which induces deviations in orientation between the portion of the fiber held in the rotation stages and the portion clamped between the load applying plates. The flattened shape of the outer cladding is therefore responsible for the flattened portions in the angular sensitivity of the fiber observed in Figure 3. The simulation nevertheless confirms that the sensitivity is maximal when the load is applied along the slow axis of the fiber. We therefore chose this orientation to embed the fibers into the composite material samples to reach the highest sensitivity in the transversal direction of the samples.

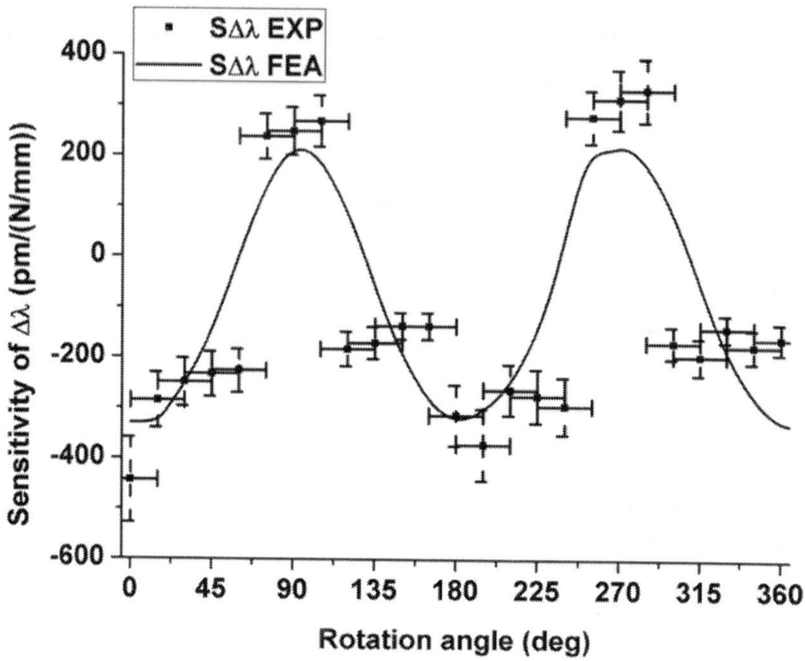

Figure 3. Comparison between experimental (EXP) and simulated values (FEA): sensitivity of the wavelength separation*versus* fiber orientation.

RESPONSE OF THE EMBEDDED FBGS FOR VARIOUS LOADING CONDITIONS

The MOFs were embedded in composite materials in the previously defined orientation. Three different types of loads were applied to the test specimens to compare the sensor responses: axial strain, transversal strain and temperature tests were carried out.

Embedding Process and Verifications

The FBGs have been embedded at mid-thickness of a lay-up of 16 carbon fiber epoxy unidirectional prepreg layers with a total thickness of 1.54 mm (M18/M55J material). The fiber coatings have been removed using an optical fiber mechanical stripper after a chemical treatment over a few centimeters at the location of the

Bragg grating. The composite samples are processed in an autoclave using the vacuum bagging technique, in which a temperature and pressure cycle is applied to the composite. The test specimens have a symmetric lay-up ($[90_2, 0_2]_{2S}$) and the optical fibers are oriented with their fast axes parallel to the surface of the composite coupon. Visual inspection of the fiber orientation (after the loading experiments) revealed a very low deviation from the desired orientation of about ±3°.

The optical fibers have much greater dimensions than the carbon reinforcement fibers of the plies. Embedding the fibers could therefore result in a geometric distortion of the surrounding fibers by creating resin-rich pockets [22]. To avoid that problem the optical fibers are integrated in the same direction as the reinforcement fibers. As seen in Figure 4(a), the optical fiber is homogeneously surrounded by resin and reinforcement fibers which should have the least detrimental consequences on the structural integrity of the host material. After embedding, the optical spectrum is still visible and both peaks are clearly distinguishable, but we notice a shift of both peaks to higher wavelengths [Figure 4(b)]. This peak shift is the result of the residual strains occurring during the production process of the laminate and arising from the mismatch in material properties between the reinforcement fibers and the resin. This was also shown in [3], where the curing process of a composite laminate was monitored by a FBG inscribed in a conventional HB optical fiber to investigate the possibility to monitor the different effects of axial and transversal stress and to determine the residual stress.

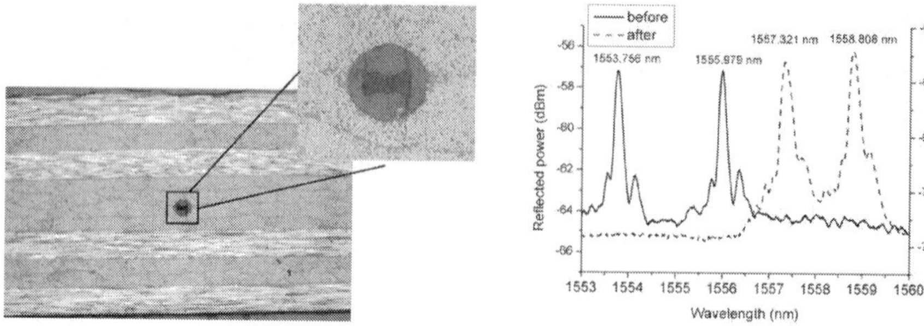

(a) (b)

Figure 4. (a) Transversal cross-section of the composite samples at the grating location. **(b)** Spectrum of the Bragg grating sensor before and after embedding in composite materials.

Temperature Tests

The samples were submitted to a thermal cycle in a dedicated climate chamber; the temperature was increased from ambient to 120 °C and then the samples were allowed to cool down slowly over a period of 10 hours. The embedded FBGs responses were monitored during the cooling phase which is considered as a quasi-static temperature variation. The temperature sensitivities of both peaks were determined by linear fits of the curves [Figure 5(b)] with R^2 coefficients above 0.994. The temperature responses of the slow and fast axes are respectively 8.02 pm/°C and 3.61 pm/°C. The different sensitivity of both modes is due to the release of the asymmetric strains occurring in the composite sample during its cooling to ambient temperature. The temperature sensitivity of the Bragg peak separation is about of 4.41 pm/°C [Figure 5(a)].

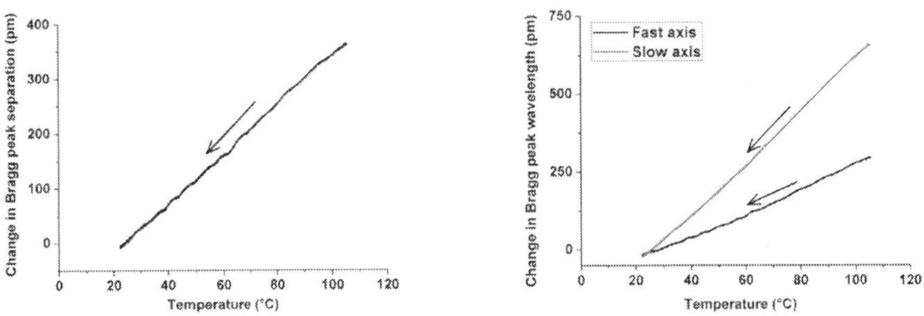

(a) (b)

Figure 5. Responses of the embedded FBGs during the cooling phase; the temperature variation is indicated with arrows. **(a)**Change in Bragg peak separation *versus* temperature. **(b)**Change in Bragg peak wavelength *versus* temperature.

The temperature sensitivity is not inherent to the fiber, but should be understood as the result of the modified stress/strain state induced by a temperature change inside the composite sample. We have indeed already demonstrated that this microstructure is intrinsically insensitive to temperature [16] and hence the wavelength changes occurring during the thermal cycles do come from strain of the surrounding material. Moreover for the bare optical fiber the Bragg peak separation sensitivity is about −0.07 pm/°C which is more than 60 times lower than the Bragg peak separation sensitivity for the FBG embedded in composite laminate. The temperature sensitivity is induced by the outer strain and therefore the proposed sensor holds potential to evaluate process-induced residual strain in composite samples and therefore to identify the initial internal state of a composite structure.

Axial Strain Test

The samples were loaded in the axial direction corresponding to the integrated optical fiber direction and the wavelength changes were monitored at the same time. The axial strain applied to the samples is measured by a reference grating glued on top of the sample at the exact location of the Bragg grating of interest [Figure 6(a)]. A maximum strain of 1,000 με is applied to the laminate at a rate of 200 με/min.

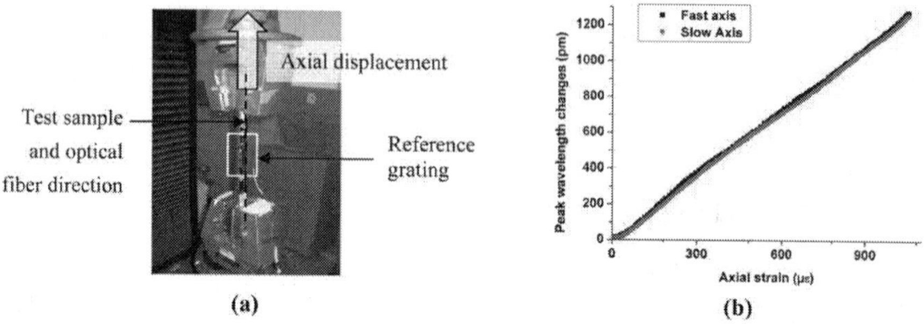

Figure 6. (a) Axial load set-up. **(b)** Peak wavelength changes*versus* axial strain.

Figure 6(b) shows that both peaks react in the same way and in a linear manner. Consequently the peak separation of the sensor is almost insensitive to axial strain for both samples (Table 1) and the fast and slow axes sensitivities are very close to those of the bare fiber (about 1.18 pm/με).

Table 1. Axial strain sensitivity of FBGs sensors embedded in $[0_2, 90_2]_{2S}$ lay-up composite.

pm/με	Average on sample tested
Fast axis	1.22 ± 0.03
Slow axis	1.21 ± 0.02
Peak separation	−0.01 ± 0.01

Transverse Strain Test

Transverse strain tests were performed on the sensors to characterize the out-of-plane laminate response. The specimens are compressed between two metal blocks with an increasing load of maximum 20 kN which corresponds to about 25 MPa (8 cm^2 contact surface). To ensure a homogenous pressure on the composite, two rubber slices are placed between the metal blocks and the load applying surfaces [Figure 7(a)]. The wavelength of the peaks and the load are recorded continuously (read-out frequency 1 Hz) during the test. By approximating the Young modulus of the laminate in the load direction by 6 GPa, we can estimate the transversal strain applied to the samples.

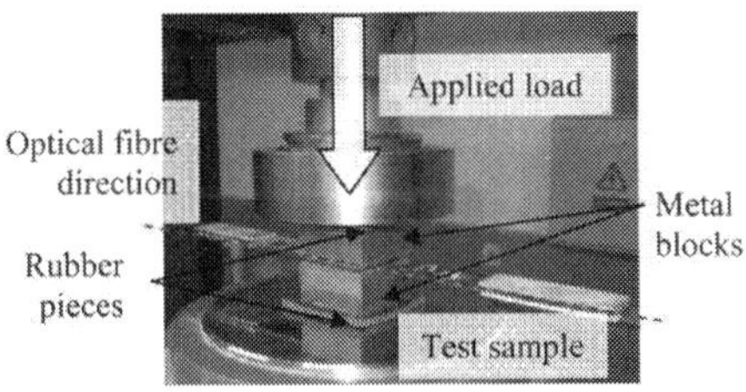

(a)

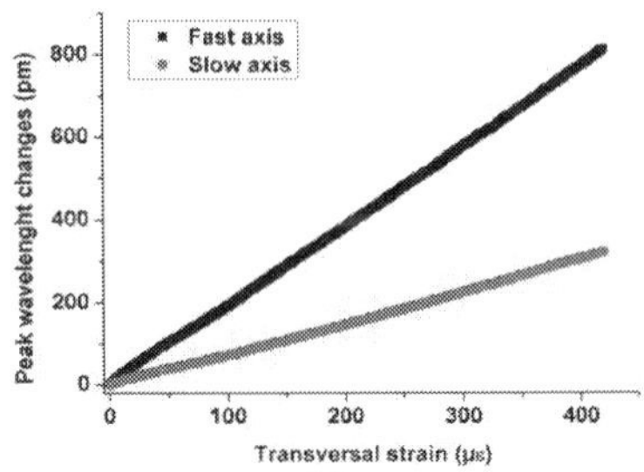

(b)

Figure 7. (a) Transverse load set-up. **(b)** Peak wavelength changes *versus* transversal strain.

Figure 7(b) shows the linear behavior of the peak wavelength *versus* the transversal strain applied, from which we determine the response of the sensors (Table 2). As for the bare fiber calibration, this experiment shows that when the optical fiber is loaded along its slow axis, its response is much more important for the fast axis than for the slow axis; in this case the difference in sensitivity is more than 50%. As shown in Figure 1(a) the fiber microstructure is highly asymmetric and presents much more side-holes along the fast axis. Therefore, the stresses along the major diameter of the cladding are larger than those created along the minor diameter direction, and hence the transverse sensitivity along the fast axis of the fiber is higher compared to the slow axis.

Table 2. Transversal strain sensitivity of the MOF.

pm/με	Average on the tested samples
Fast axis	0.26 ± 0.05
Slow axis	0.10 ± 0.02
Peak separation	−0.16 ± 0.03

The specific behavior of the sensors can hence be understood as follows. When the fiber is non-uniformly loaded at constant temperature, one can de-multiplex the loads acting on the optical fiber:

- a Bragg peak separation change is induced by transversal strains, because the axial strain sensitivity of the peak separation has been shown to be zero;
- identical changes in the peak wavelengths will refer primarily to axial strains. The sensitivities of the slow and fast axis for axial strain are 5 to 10 times larger than for transversal strain.

We can now assess in how much more sensitive our sensor is compared to FBGs in conventional HB optical fibers and in a MOF with another type of microstructure.

Comparison of the Sensor Response with Previous Studies

In [23] the transversal strain response of sensors was determined for conventional bow-tie type HB optical fiber and for another MOF structure (referred to here as MOF2) both embedded in a unidirectional laminate. Figure 8 shows that the proposed MOF (referred to as MOF1) demonstrates a much higher response to transversal strain than the two other FBGs. The transversal Bragg separation sensitivity reaches about −0.16 pm/με in the case of the

MOF studied here; which is one order of magnitude higher than that of the conventional HB fiber and of MOF2. However the laminate composite in [23] refers to a unidirectional configuration and one therefore needs to verify how large the composite configuration influences the sensor response.

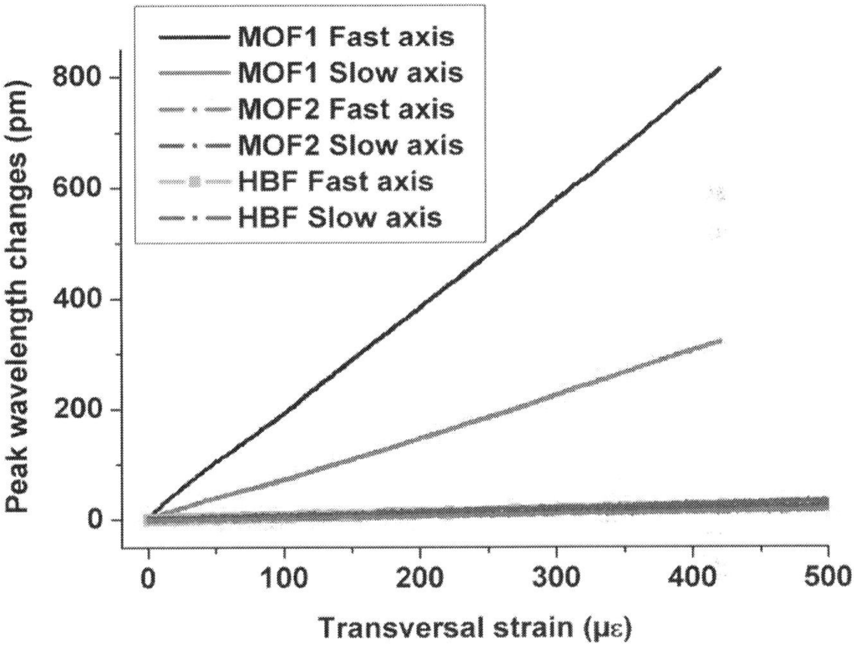

Figure 8. Transversal strain response of a conventional HB fiber of bow-tie type (HBF) and of the MOF2 embedded in unidirectional laminate compared with the response of MOF1 integrated in a cross-ply configuration (the sensitivity plots of MOF2 and HBF are overlaying each other).

To determine how important the laminate configuration is we perform numerical simulations of the transversal strain tests. We first validate the simulations by comparing the numerical and experimental response of MOF1 and then we model the behavior of MOF2 embedded in the cross-ply configuration.

The modeling of the laminate behavior submitted to transversal pressure is carried out with the commercially available finite element software ABAQUS. The goal is to link the strain in the fiber core to the strain in the laminate composite using Equations (4) and (5). The strain values used in these equations are an average of the principal strains in the entire core region all along the grating length; averaging the strain allows avoiding stress non-uniformity coming from the air-hole microstructure. We assume in the analysis that the principal strain axes and the optical axes of the anisotropic fiber overlap. We assume that the bonding between the optical fiber and the host material is perfect, the deformations are considered small and both materials behave elastically. The composite laminate model, illustrated inFigure 9(a), consists of 100,500 linear hexahedral elements of type C3D8R. The mesh is refined in the vicinity of the optical fiber. For symmetry reasons, only one eighth of the specimen is represented and specific boundary conditions are applied. The top surface is submitted to a linearly increased load of maximum 25 MPa. The simulated length is 20 mm which ensures us to be far away from the end facets and hence we avoid boundary effects.

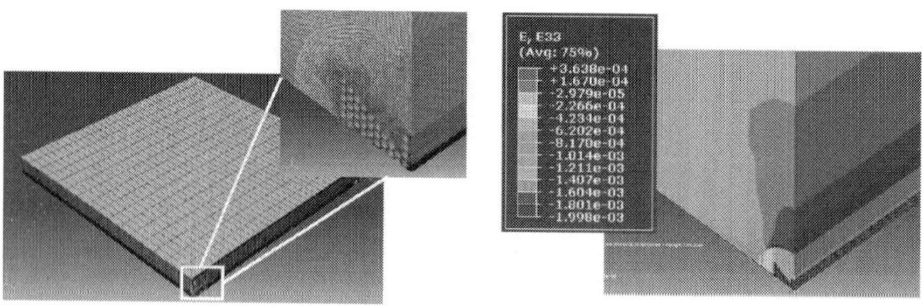

(a) (b)

Figure 9. (a) 1/8 of the laminate structure meshed and close-up of MOF1. **(b)** Strain in the out-of-plane direction for MOF1.

Figure 10 summarizes the results of the finite element analysis (FEA). Figure 10(a) first evidences the excellent agreement between simulations and experiments for MOF1 embedded in a cross-ply laminated composite, with a relative error of only 4.5 %. This validates our model and FEA approach. Second, the model suggests that when the same MOF is embedded within a unidirectional laminated composite the sensitivity to applied load is slightly reduced by approximately 3.4% [Figure 10(a)]. This result indicates that the composite lay-up does not play a major role in the transversal sensor response. Finally, the simulations again reveal the substantially improved sensitivity of MOF1 with a factor of 6 compared to MOF2 when both fiber types are embedded within a composite with similar lay-up [Figure 10(b)].

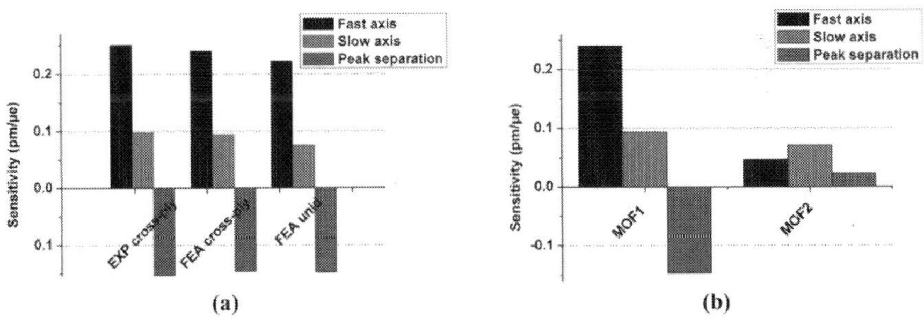

Figure 10. Comparison between the composite configurations. **(a)** MOF1 embedded in cross-ply laminate (experimental (EXP) and modeling (FEA)) and unidirectional laminate (modeling). **(b)** MOF1 and MOF2 in cross-ply laminate (modeling results).

CONCLUSIONS

We have embedded fiber Bragg gratings written in specially designed highly birefringent microstructured optical fiber in laminate fiber reinforced composite material samples, with the intention to accurately monitor the transverse strain components inside composite materials. The sensor response is encoded in the spectral distance between the two reflection peaks of the fiber Bragg grating. The gratings were successfully integrated within the composite panels without compromising the mechanical integrity of the material and the optical properties of the grating. The

14. Bertholds, A; Dandliker, R. Determination of the individual strain-optic coefficients in single mode optical fibers. *J. Lightwave Technol* **1988**, *6*, 17–20.
15. Russel, PSTJ. Photonic-crystal fibers. *J. Lightwave Technol* **2006**, *24*, 4729–4749.
16. Martynkien, T; Statkiewicz-Barabach, G; Olszewski, J; Wojcik, J; Mergo, P; Geernaert, T; Sonnenfeld, C; Anuszkiewicz, A; Szczurowski, MK; Tarnowski, K; Makara, M; Skorupski, K; Klimek, J; Poturaj, K; Urbanczyk, W; Nasilowski, T; Berghmans, F; Thienpont, H. Highly birefringent microstructured fibers with enhanced sensitivity to hydrostatic pressure. *Opt. Express***2010**, *14*, 15113–15121.
17. Rothhardt, MW; Chojetzki, C; Mueller, HR. High-mechanical-strength single-pulse draw tower gratings. Proceedings of SPIE Photonics North 2004, Ottawa, Canada, 2004; 5579, pp. 127–135.
18. Geernaert, T; Becker, M; Nasilowski, T; Wojcik, J; Urbanczyk, W; Rothhardt, M; Bartelt, H; Terryn, H; Berghmans, F; Thienpont, H. Bragg grating inscription in GeO_2-doped microstructured optical fibers. *IEEE J. Lightwave Technol* **2010**, *28*, 1459–1467.
19. Chehura, E; Ye, CC; Staines, SE; James, SW; Tatam, RP. Characterization of the response of fibre Bragg gratings fabricated in stress and geometrically induced high birefringence fibres to temperature and transverse load. *Smart Mater. Struct* **2004**, *13*, 888–895.
20. Geernaert, T; Luyckx, G; Voet, E; Nasilowski, T; Chah, K; Becker, M; Bartelt, H; Urbanczyk, W; Wojcik, J; De Waele, W; Degrieck, J; Terryn, H; Berghmans, F; Thienpont, H. Transversal load sensing with fiber Bragg gratings in microstructured optical fibers. *IEEE Photonics Technol. Lett***2009**, *21*, 6–8.
21. Szpulak, M; Martynkien, T; Urbanczyk, W. Highly birefringent photonic crystal fibre with enhanced sensitivity to hydrostatic pressure. Proceedings of the 8th International Conference on Transparent Optical Networks, Nottingham, UK, June 2006; 4, pp. 174–177.
22. Huang, CY; Trask, RS; Bond, IP. Characterization and analysis of carbon fibre-reinforced polymer composite laminates with embedded circular vasculature. *J. R. Soc. Interface* **2010**, *7*, 1229–1241.

23. Luyckx, G; Voet, E; Geernaert, T; Chah, K; Nasilowski, T; De Waele, W; Van Paepegem, W; Becker, M; Bartelt, H; Urbanczyk, W; Wojcik, J; Degrieck, J; Berghmans, F; Thienpont, H. Response of FBGs in microstructured and bow tie fibers embedded in laminated composite. *IEEE Photonics Technol. Lett* **2009**, *21*, 1290–1292.

CITATION

Camille Sonnenfeld, Sanne Sulejmani, Thomas Geernaert, Sophie Eve, Nicolas Lammens, Geert Luyck, Eli Voet, Joris Degrieck, Waclaw Urbanczyk, Pawel Mergo, Martin Becker, Hartmut Bartelt, Francis Berghmans and Hugo Thienpont, Microstructured Optical Fiber Sensors Embedded in a Laminate Composite for Smart Material Applications, doi:10.3390/s110302566.

CHAPTER 9

Design of an Embedded CMOS Temperature Sensor For Passive RFID Tag Chips

Fangming Deng[1,2,*], *Yigang He*[2], *Bing Li*[2], *Lihua Zhang*[1], *Xiang Wu*[1], *Zhihui Fu*[1] *and Lei Zuo*[2]

[1]School of Electrical and Electronic Engineering, East China Jiaotong University, Nanchang 330013, China; E-Mails: lhzhang@ecjtu.edu.cn (L.Z.); zgxiangyu@163.com (X.W.); fuzhihui @sohu.com(Z.F.)

[2]School of Electrical Engineering and Automation, Hefei University of Technology, Hefei 230009, China; E-Mails: 18655136887@163.com (Y.H.); libinghnu@163.com (B.L.); 186551 18171@126.com(L.Z.)

ABSTRACT

This paper presents an ultra-low embedded power temperature sensor for passive RFID tags. The temperature sensor converts the temperature variation to a PTAT current, which is then transformed into a temperature-controlled frequency. A phase locked loop (PLL)-based sensor interface is employed to directly convert this temperature-controlled frequency into a corresponding digital output without an external reference clock. The fabricated sensor occupies an area of 0.021 mm² using the TSMC 0.18 1P6M mixed-signal CMOS process. Measurement results of the embedded sensor within the tag system shows a 92 nW power dissipation under 1.0 V supply voltage at room temperature, with a sensing resolution of

0.15 °C/LSB and a sensing accuracy of −0.7/0.6 °C from −30 °C to 70 °C after 1-point calibration at 30 °C.

INTRODUCTION

Radio Frequency Identification (RFID) and Wireless Sensor Networks (WSN) are the two key technologies of the Internet of Things. RFID is a technology to automatically identify objects or humans by utilizing radio waves and WSN is a technology to cooperate in network for monitoring environmental conditions by utilizing various sensor nodes. Recently research on the integration of RFID and WSN has been a hot topic [1]. RFID can extend the ability of a WSN by providing sensible properties to otherwise unsensible objects. WSN, on the other hand, being able to monitor physical events, can provide much more information for the measurement of temperature, humidity, pressure, *etc.*, than simple RFID. Moreover, in an RFID system, reader-tag communications are conducted in single hops without inter-communication among tags. RFID systems integrated with WSNs will enjoy the advantages of wireless multi-hop inter-communication over wider areas [2].

Integrating a RFID tag with sensor nodes is the main type of integration of RFID and WSNs. The passive RFID tag offers several advantages such as battery-less operation, high flexibility, low cost and fast deployment, which all result in its extensive applications in commercial use [3]. The passive RFID tag collects the wireless energy from the RFID reader as its power supply. Hence, the power dissipation performance, determining the maximum reading distance of the RFID reader, is crucial for the design of a passive RFID tag and its integrated sensor. Secondly, the occupied sensor area and its calibration effort should be minimized to maintain low cost. Thirdly, as the sensor is supplied by the on-chip power management unit, a good supply rejection is necessary to achieve accurate sensing. In recent years RFID tags integrated with temperature sensors have already introduced in several papers [4,5].

The traditional CMOS temperature sensors are based on the Bipolar Junction Transistor (BJT) [6,7,8,9]. These sensors convert temperature variation into a corresponding voltage, which is followed by an Analog to Digital Converter (ADC) to generate a digital output. However such sensors show power consumption in the mW range, which is unsuitable for passive RFID applications. Hence MOS-based temperature sensors targeted for wireless system have been introduced. For low power operation, time-to-digital [4,10,11,12] or frequency-to-digital conversion [5,13,14] is used instead of ADCs. The sensors consume less power than BJT-based sensors at the expense of resolution and accuracy. However, these sensors need an external clock as a reference, which deteriorates the power and accuracy performance of the sensors.

This paper presents a novel CMOS temperature sensor for RFID applications. The proposed temperature sensor employs a Phase-Locked Loop (PLL) architecture and can transfer the sensor values to a digital output in the frequency domain without a reference clock. The rest of the paper is organized as follows: Section 2 first introduces the PLL-based sensor architecture and then illustrates the circuit design of the temperature sensor in detail. The measurement results are presented and compared with previous designs in Section 3. Finally, some conclusions are drawn in Section 4.

PLL-BASED TEMPERATURE SENSOR ARCHITECTURE

The proposed temperature sensor theory, developed from the PLL-based interface theory [15,16], is shown in Figure 1a. It consists of two main blocks: a frequency-modulating block, which converts the temperature information into the frequency domain, and a frequency-demodulating block, which converts the frequency into the digital domain, resulting in a complete temperature-to-digital flow. The frequency-modulating block consists of a Temperature-Controlled Oscillator (TCO) that directly converts the temperature information into a corresponding frequency f_t. The frequency-demodulating block is a digital first-order PLL, consisting of a 1-bit Phase Detector (PD) and a Digitally-Controlled Oscillator (DCO). This PLL measures whether f_t leads or lags f_d (the output frequency of the DCO) and hence the DCO is only steered by a 1-bit signal b_o. When the entire feedback loop is locked, f_d shifts between a maximum and minimum value (f_{max} and f_{min}) which correspond to the maximal and minimal value of the temperature (seen in Figure 1b). The average value of f_d will equal f_t. Therefore, the over-sampled outputb_o represents the digital value of the temprature information.

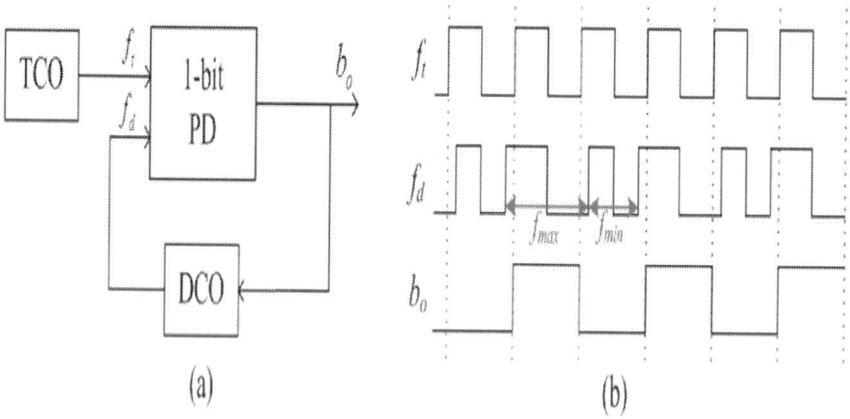

Figure 1. PLL-based temperature sensor; **(a)** architecture; **(b)** corresponding waveforms in locked state.

The difference between these two frequencies defines the lock range of this first-order PLL. Rearrangement of the building blocks shows that this structure is similar to a single-bit first-order noise-shaped ∑Δ-modulator, but operating in the frequency domain instead of the voltage domain. The first-order PLL can be seen as a frequency-to-digital ∑Δ-modulator in which the frequency error is integrated and single-bit quantized, which means that the quantization noise is first-order noise-shaped to the output [17].

The proposed temperature sensor architecture, based on the PLL theory, is shown in Figure 2. Both the TCO and the DCO are implemented as N-stage inverter-based ring oscillators. The oscillating frequencies of TCO and DCO depend on the time it takes a transition to propagate around the loop. For a standard ring oscillator and assuming an equal rise and fall time for the different stages respectively, the oscillating frequency f_{osc} of the ring oscillator can be expressed as follows:

$$f_{osc} = \frac{1}{t_d} = \frac{I_l}{V_m C_l} \tag{1}$$

where t_d is the delay time of the loop, I_l is the current flowing through the inverter, C_l is the total output capacitor of the loop, and V_m is the swing range of the output voltage which mostly equals to the supply voltage V_{DD}. Because each inverter will add extra capacitance and delay time, the stage number N is determined by the frequency f_{osc}.

In this design, the TCO and DCO inverters have identical schematics except for their current sources and adopt a current-starved architecture for lower power dissipation and higher temperature stability. According to Equation (2), if the current source of the TCO I_t is a Proportional To Absolute Temperature (PTAT) current, then the oscillating frequency of the TCO f_t is a PTAT frequency. The current source $I_s$of the DCO is a constant current. Within the measured temperature range, the f_t can be considered as consisting of two parts: the quiescent frequency f_{to} and the variable frequency f_{tm}. The DCO is steered by the PD output signal b_o, which is a representation of the phase difference between the TCO and the DCO. The variable capacitive load on a single stage of the DCO consists of a switch and a capacitor C_m. The capacitor C_m is swapped in or out of the DCO depending on the output b_o from the PD. When C_m is disconnected from the DCO, the oscillating frequency of the DCO f_d is designed to be equal to f_{to}. WhenC_m is connected to the DCO, f_d is designed to be slightly larger than f_t. The PD is simply composed of a 1-bit d-flip-flop. When the entire feedback loop is locked, the average value of f_d will correspond to the f_t. Therefore, the b_o represents the digital value of the corresponding temperature.

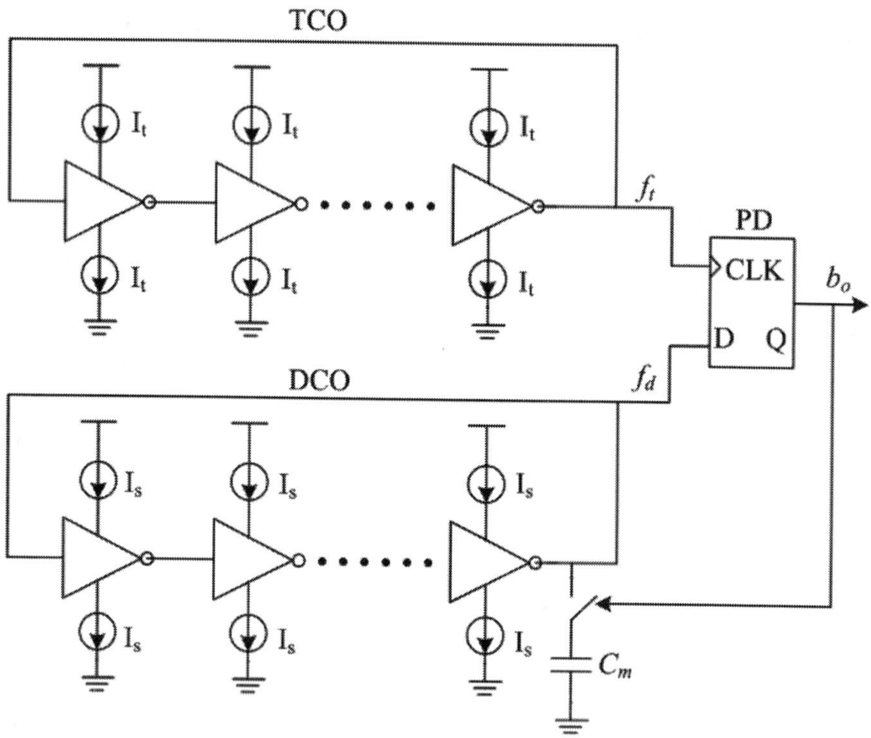

Figure 2. Proposed temperature sensor architecture.

This proposed temperature sensor has a simple architecture and can adopt fully-digital blocks, resulting in low power dissipation and low chip area. Furthermore, this architecture can convert the temperature information into digital signals without an extra reference clock, resulting in improved accuracy performance. The conversion is based on locking two oscillators, which means that the relative frequency is of importance and not the absolute frequency. Hence any influence such as supply voltage and temperature variations will be canceled out if the two oscillators are completely matched. The matching between the TCO and DCO plays an important role in this design. Mismatch between the TCO and DCO can be expressed as an offset between the coefficients of the polynomials of the characteristics of the oscillators, resulting in a deterioration of the Effective Number Of Bits (ENOB). Considering the issues of system linearity and process variation, C_m is normally designed as a programmable capacitor.

PROPOSED TEMPERATURE SENSOR CIRCUIT DESIGN

The key design of the proposed temperature sensor in Figure 3 is to generate a PTAT current source I_t. Inspired by Amaravati [18], we design a Process and Voltage (PV)-compensated PTAT current source as shown in Figure 3. The transistors M_1 and M_2 are used to produce a PTAT voltage V_{PTAT}, which is then applied on a PV-compensated resistor to produce the PV-compensated current I_t. The transistors M_3–$M_5$are used to produce this PV-compensated active resistor.

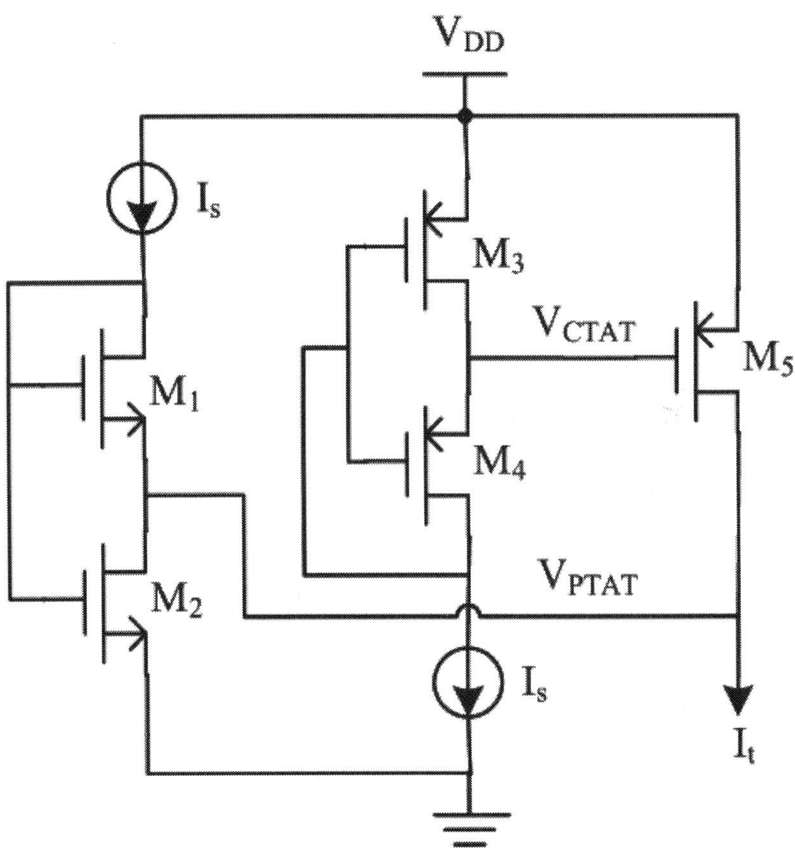

Figure 3. Proposed PTAT current source.

The NMOS transistors M_1 and M_2, biased in the subthreshold region, are short channel and long channel transistors respectively. The PTAT voltage V_{PTAT} can be obtained as follows [19]:

$$V_{PTAT} = V_{GS1} - V_{GS2} = \eta V_T \ln K \tag{2}$$

where V_{GS1} and V_{GS2} are the gate-source voltages of the M_1 and M_2 respectively, η is the subthreshold slope factor with an approximate value of 1.45, V_T is the thermal voltage and K is the width ratio of this two transistors. The PMOS transistor M_5 works in the triode region as an active resistor, which resistance can be expressed by:

$$R = \frac{1}{\mu_p C_{ox} (W/L)(V_{SG} - |V_{th}|)} \tag{3}$$

where μ_p is the hole mobility, C_{ox} is the oxide capacitance of the transistor, W/L is the ratio of the width respect to the length of the transistor, V_{SG} is the source-gate voltage of the transistor and V_{th} is the threshold voltage. The effect of temperature variation of on μ_p the required gate voltage is larger than that of V_{th}. Hence we need a gate voltage with Complementary To Absolute Temperature (CTAT) behavior to compensate the effect of temperature variation. The PMOS transistors M_3 and M_4 are biased in the subthreshold region and the voltage V_{CTAT} can be expressed as follows:

$$V_{CTAT} = V_{DD} - (V_{SG4} - V_{SG3}) = V_{DD} - \eta V_T \ln K \tag{4}$$

Equation (5) shows that the produced voltage is a CTAT voltage which also tracks supply voltage variations. Hence the output current I_t is the PV-compensated PTAT current source. The simulation results of the proposed PTAT current source on different process corners are shown in Figure 4. The I_t achieves good linearity with temperature and the simulated worst case variation across corners is ±4.5%.

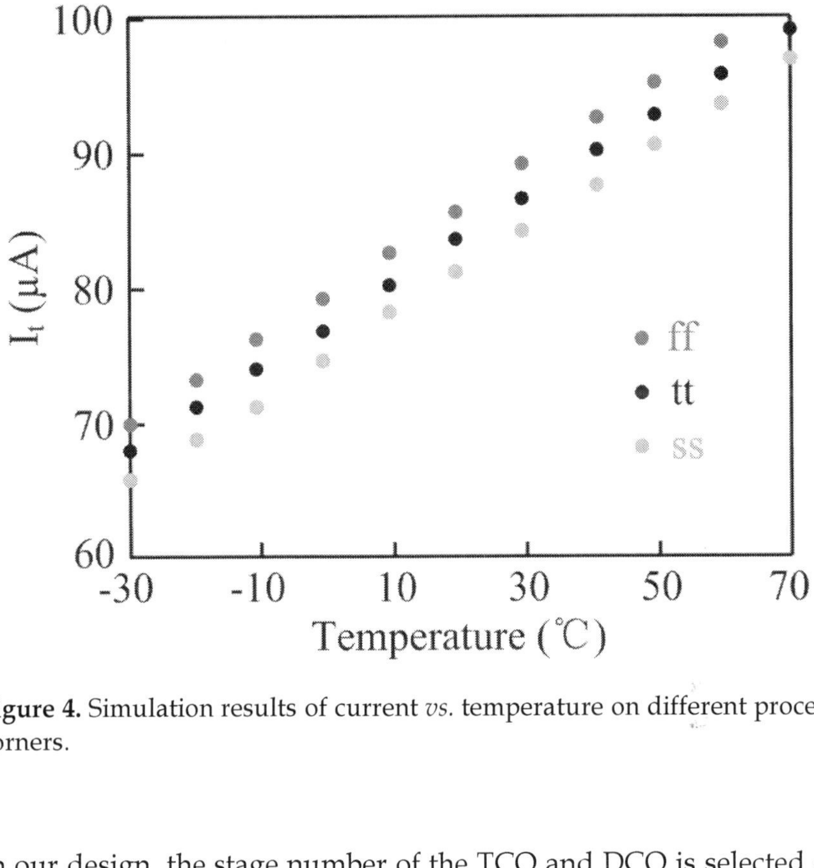

Figure 4. Simulation results of current *vs.* temperature on different process corners.

In our design, the stage number of the TCO and DCO is selected as 3. In a conventional ring oscillator (shown in Figure 5a), the transistors M_1–M_6 form the three-stage inverters and the transistors M_7–M_{12} form the current mirrors. The output nodes of all inverter cells experience a rail-to-rail swing from ground to the supply voltage V_{DD}. However in the proposed method (shown in Figure 5b) [20], four extra transistors (M_{H1}, M_{H2}, M_{L1}, M_{L2}) are employed to constrain the voltage swing of internal nodes for reducing the overall power consumption, while the output node is left untouched to vary between the supply voltage and ground [21].

The four extra transistors (M_{H1}, M_{H2}, M_{L1}, M_{L2}) operate as follows: when the input switches from high-to-low, the inverter's output starts to rise from low-to-high. As the output voltage rises, the gate voltages of M_{H1} and M_{L2} increase, which results in lowering the gate-source voltage of M_{H1} and M_{L2}, V_{GSH}. When V_{GSH} falls below the threshold voltage and then M_{H1} and M_{L2} enter the subthreshold region where the current decreases exponentially. Therefore we can conclude the voltage headroom of the internal inverters is reduced to

$$V_{DD} - (V_{GSH} + V_{GSL}) - (V_{DSH} + V_{DSL}) \tag{5}$$

where V_{DD} is the supply voltage, V_{GSL} is the gate-source voltage of M_{L1} and M_{L2}, V_{DSH} is the drain–source voltage of M_7 and M_8 and V_{DSL} is the drain–source voltage of M_{10} and M_{11}.

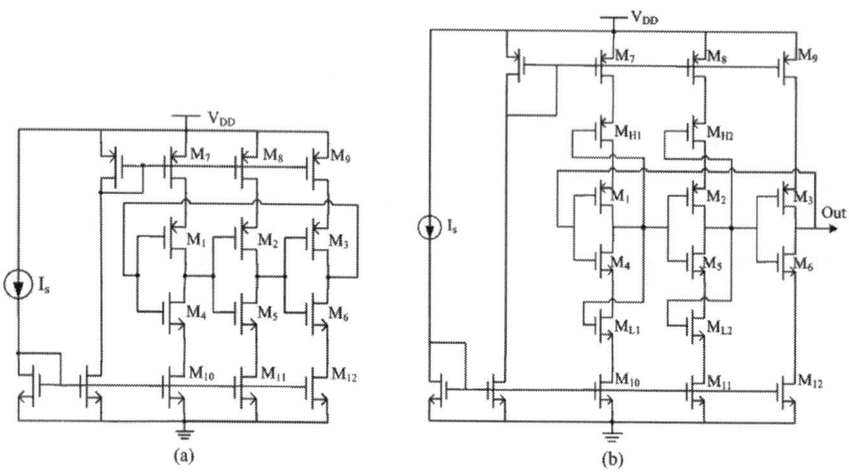

Figure 5. Current-starved ring oscillator; (**a**) conventional schematic; (**b**) proposed schematic.

The simulation results, shown in Figure 6, indicate that the output node swings between the supply voltage 1.0 V and the ground, while the internal nodes vary from 0.3 V to 0.7 V. The reduction of the voltage swing to nearly 50% of its nominal value lowers the dynamic power consumption of the internal nodes by 75% and the overall power consumption by much more than 25%.

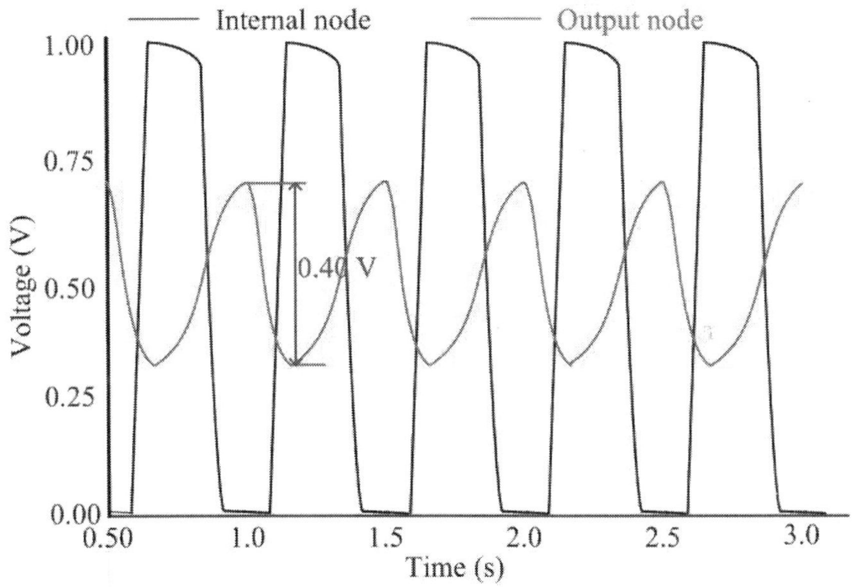

Figure 6. Output waveforms of the proposed ring oscillator at different nodes.

MEASUREMENT RESULTS

The RFID tag with the proposed temperature sensor was implemented in the Taiwan Semiconductor Manufacturing Company (TSMC) 0.18 μm 1P6M mixed-signal CMOS process. As shown in Figure 7a, the RFID tag chip covers an area of 1.75 × 1.6 mm^2 without pad and the temperature sensor occupies an active area of 0.021 mm^2. Extensive matching is exercised during the layout design stage in order to minimize the effects due to process variation and mismatch. In order to characterize the performance of the temperature sensor, the measurement is performed inside a VCL 4000 temperature chamber from Votsch Industry Electronics (Taichang, China), together with a calibrated thermometer as a reference. The output of the sensor tag is compared with the reference thermometer to analyze the linearity. The wireless test environment consists of a special RFID test instrument, a VISN-R1200 from VI Service Network (Shanghai, China) and an anechoic box (seen in Figure 7b). The VISN-R1200 can work as an adjustable RFID reader and display the testing signals simultaneously. The RFID tag was tested in the anechoic box for the electromagnetic shielding.

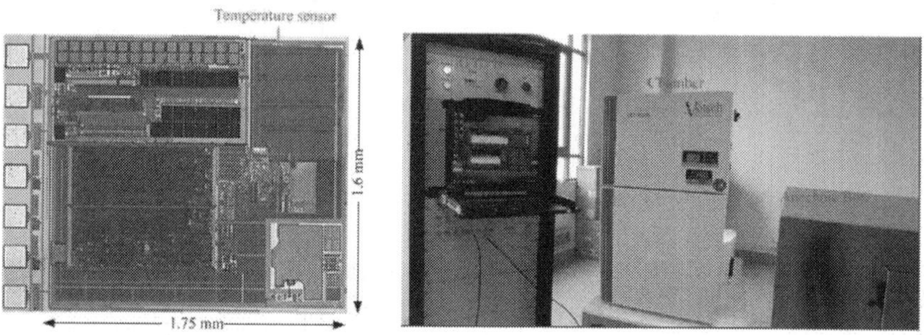

(a) (b)

Figure 7. (**a**) Chip micrograph of the complete RFID tag; (**b**) Measurement equipments.

Figure 8a shows the measured different outputs of the temperature sensor at −5 °C and 25 °C. It is easily concluded that the proposed temperature sensor shows higher oscillation frequency and average output duty cycle at 25 °C.

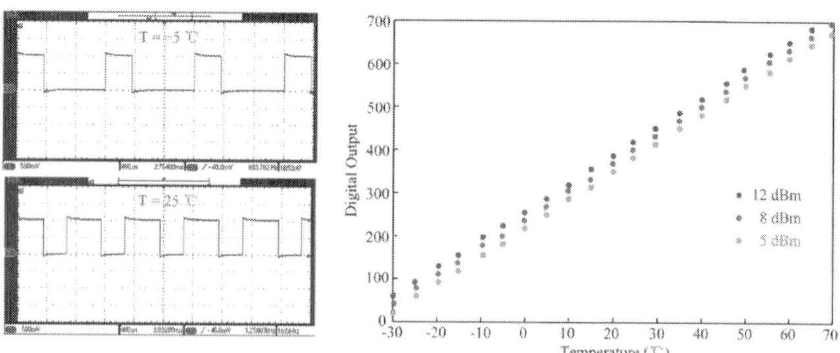

(a) (b)

Figure 8. Performances of the test temperature sensor; (**a**) corresponding outputs of the sensor interface at −5 °C and 25 °C; (**b**) digital output of baseband *vs.* temperature.

The output of the temperature sensor is further processed by the digital baseband block in the RFID tag. The relation between the digital output of baseband and the temperature variation is shown in Figure 8b. The sensor tag was connected to the signal generator output and was tested under three different input power levels: 12, 8 and 5 dBm, respectively. The test sensor achieves about 0.15 °C /LSB resolution with a 1 kHz conversion frequency and high linearity performance within the measured range from −30 °C to 70 °C. The error of the linearity under different input power is small, which proves that the proposed temperature sensor has good immunity to supply voltage variation.

Because the sensor tag needs RF power to activate the internal building blocks, input power variation might affect the temperature sensor performance. Figure 9a shows the measured errors of two sensor tags under different input RF power levels. When the input power is small (≤5 dBm), the power received by the tags is not adequate to activate the tags, leading to the obvious sensing error that is observed. By increasing the input power, the sensor output becomes stable from 5 dBm to 12 dBm. If input power is too high (≥12 dBm), the internal power-limiter will convert the excess power into heat resulting in obvious errors. Figure 9b gives the measured temperature sensor inaccuracy over five different test chips. After 1-point calibration at 30 °C by adjusting the capacitor C_m, the measured error is −0.7/0.6 °C from −30 °C to 70 °C.

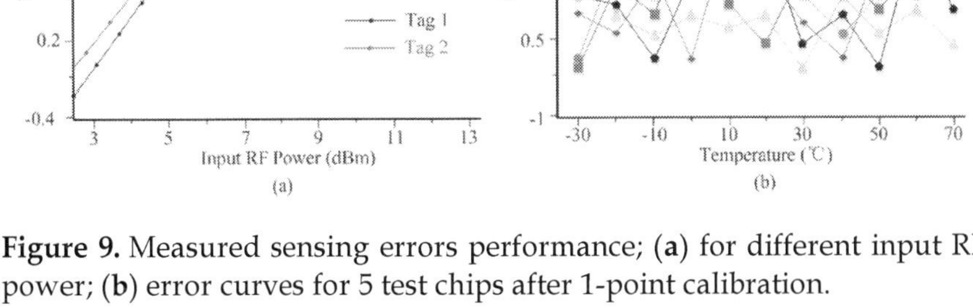

Figure 9. Measured sensing errors performance; (**a**) for different input RF power; (**b**) error curves for 5 test chips after 1-point calibration.

Table 1 compares the measured performances of the proposed design wither other state-of-the-art designs, which have different performances and different target applications. The designs in [7,9] achieve high resolution and error performance, but consume too much power, thus resulting unsuitable for passive RFID applications. The designs in [11,12,13] achieve ultra-low power dissipation at the expense of resolution and error performance. Though the design in [14] obtains excellent power and resolution performances, the operating temperature range is small. The measurement results show the proposed temperature sensor covers 0.021 mm^2 and draws 92 nA current from 1 V supply voltage. Furthermore, due to the immunity to supply voltage variation, this work requires only 1-point calibration, which ensures low-cost applications.

Table 1. Performances comparison of temperature sensor.

Reported Sensor	Process (μm)	V-Supply (V)	Power (nW)	Area (mm^2)	Resolution (°C/LSB)	Range (°C)	Error (°C)
This work	0.18	1.0	92	0.021	0.15	-30~70	-0.7/0.6
[7]	0.065	1.2	10000	0.100	0.03	-70~125	-0.2/0.2
[9]	0.16	1.5	5100	0.080	0.02	-55~125	-0.15/0.15
[11]	0.18	0.5,1	119	0.042	0.2	-10~30	-0.8/1
[12]	0.18	1.0	405	0.032	0.3	0~100	-0.8/1
[13]	0.18	1.8	900	0.200	0.5	27~47	-1~1
[14]	0.35	1.4,2.1	110	0.084	0.04	35~45	-0.1~0.1

CONCLUSIONS

An embedded temperature sensor in a passive RFID tag is designed, measured and demonstrated in this paper. The temperature sensing part of the sensor converts the temperature variation into a PTAT current, which is then transformed into a temperature-controlled frequency. The sensor interface, based on a PLL architecture, can transform the temperature-controlled frequency into the corresponding digital output directly without an external reference clock. The fabricated sensor occupies an area of 0.021 mm^2 using the TSMC 0.18 1P6M mixed-signal CMOS process. The measurement results show that the proposed sensor achieves 92 nW power dissipation with an error −0.5/0.6 °C from −30 °C to 70 °C after 1-point calibration. We will work on improving the resolution of this design in future.

ACKNOWLEDGMENTS

This work was supported by the National Natural Science Funds of China for Distinguished Young Scholar under Grant (50925727), the Key Grant Project of Chinese Ministry of Education (313018), the Young Scientists Fund of the National Natural Science Foundation of China (61401139), the Key Science and Technology Project of Anhui Province of China (1301022036) and the East China Jiaotong University Foundation (14DQ08). The authors would also like to thank TSMC semiconductor for its tape-out support and the RFID group in School of Electrical Engineering and Automation, Hefei University of Technology for their technical support.

Authors Contributions

Yigang He provides the instructions for system design and financial support for this research. Lihua Zhang, Xiang Wu and Zhihui Fu provide helps in designing the temperature sensor. Bing Li and Lei Zuo help to achieve the measurement of the sensor. All authors provide the helps in revisions of this manuscript.

Conflicts of Interest

The authors declare no conflict of interest.

REFERENCES

1. Mirshahi, S.; Uysal, S.; Akbari, A. Integration of RFID and WSN for supply chain intelligence system. In Proceedings of the 2013 International Conference on Electronics, Computers and Artificial Intelligence (ECAI), Pitesti, Romania, 27–29 June 2013; pp. 1–6.
2. Al-Turjman, F.M.; Al-Fagih, A.E.; Hassanein, H.S. A novel cost-effective architecture and deployment strategy for integrated RFID and WSN systems. In Proceedings of the 2012 International Conference on Computing, Networking and Communications (ICNC), Maui, HI, USA, 1–2 February 2012; pp. 835–839.
3. Deng, F.M.; He, Y.G.; Zhang, C.L.; Feng, W. A CMOS humidity sensor for passive RFID sensing applications. *Sensors* **2014**, *14*, 8728–8739.
4. Yin, J.; Yi, J.; Law, M.K.; Ling, Y.; Lee, M.C.; Ng, K.P.; Yuen, M. A system-on-chip EPC Gen-2 passive UHF RFID tag with embedded temperature sensor. *IEEE J. Solid State Circuits* **2010**, *45*, 2404–2420. [Google Scholar]
5. Qi, Z.W.; Zhuang, Y.Q.; Li, X.M.; Liu, W.F.; Du, Y.Q.; Wang, B. Full passive UHF RFID Tag with an ultra-low power, small area, high resolution temperature sensor suitable for environment monitoring. *Microelectron. J.* **2014**, *45*, 126–131.
6. Aita, A.L.; Pertijs, M.A.; Makinwa, K.A.; Huijsing, J.H. A CMOS smart temperature sensor with a batch-calibrated inaccuracy of ±0.25 °C (3σ) from −70 °C to 130 °C. In Proceedings of the 2009 IEEE Solid-State Circuits Conference-Digest (ISSCC), San Francisco, CA, USA, 8–12 February 2009; pp. 342–343.
7. Sebastiano, F.; Breems, L.J.; Makinwa, K.A.; Drago, S.; Leenaerts, D.M.; Nauta, B. A 1.2-V 10-μW NPN-based temperature sensor in 65-nm CMOS with an inaccuracy of 0.2 °C (3σ) from 70 °C to 125 °C. *IEEE J. Solid State Circuits* **2010**, *45*, 2591–2601.
8. Souri, K.; Makinwa, K.A.A. A 0.12 mm^2 7.4 μW micropower temperature sensor with an inaccuracy of 0.2 °C (3σ) from 30 °C to 125 °C. *IEEE J. Solid State Circuits* **2011**, *46*, 1693–1700.
9. Souri, K.; Chae, Y.; Makinwa, K.A.A. A CMOS temperature sensor with a voltage-calibrated inaccuracy of 0.15 C (3σ) from 55 °C to 125 °C. *IEEE J. Solid State Circuits* **2013**, *48*, 292–301.

10. Wang, B.; Law, M.K.; Bermak, A.; Luong, H.C. A passive RFID tag embedded temperature sensor with improved process spreads immunity for a −30 °C to 60 °C sensing range. *IEEE Trans. Circuits Syst. I Regul. Pap.* **2014**, *61*, 337–346.
11. Law, M.K.; Bermak, A.; Luong, H.C. A sub-μW embedded CMOS temperature sensor for RFID food monitoring application. *IEEE J. Solid State Circuits* **2010**, *45*, 1246–1255.
12. Law, M.K.; Bermak, A. A 405-nW CMOS temperature sensor based on linear MOS operation.*IEEE Trans. Circuits Syst. II Express Briefs* **2009**, *56*, 891–895.
13. Zhou, S.; Wu, N. A novel ultra-low power temperature sensor for UHF RFID tag chip. In Proceedings of the 2007 IEEE Asian Solid-State Circuits Conference (ASSCC), Jeju, Korea, 12–14 November 2007; pp. 464–467.
14. Vaz, A.; Ubarretxena, A.; Zalbide, I.; Pardo, D.; Solar, H.; Garcia-Alonso, A.; Berenguer, R. Full passive UHF tag with a temperature sensor suitable for human body temperature monitoring. *IEEE Trans. Circuits Syst. II Express Briefs* **2010**, *57*, 95–99.
15. Danneels, H.; Piette, F.; de Smedt, V.; Dehaene, W.; Gielen, G. A novel PLL-based frequency-to-digital conversion mechanism for sensor interfaces. *Sens. Actuators A Phys.* **2011**, *172*, 220–227.
16. Danneels, H.; Piette, F.; Gielen, G. A novel PLL-based sensor interface for resistive pressure sensors. *Procedia Eng.* **2010**, *5*, 62–65.
17. Van Rethy, J.; Danneels, H.; Gielen, G. Performance Analysis of Energy-Efficient BBPLL-Based Sensor-to-Digital Converters. *IEEE Trans. Circuits Syst. I Regul. Pap.* **2013**, *11*, 1881–1886. [Google Scholar]
18. Amaravati, A.; Dave, M.; Baghini, M.S.; Sharma, D.K. 800-nA process-and-voltage-invariant 106-dB PSRR PTAT current reference. *IEEE Trans. Circuits Syst. II Express Briefs* **2013**, *60*, 577–581.
19. Rossi, C.; Galup-Montoro, C.; Schneider, M.C. PTAT voltage generator based on an MOS voltage divider. In Proceedings of the 2007 NSTI, Shanghai, China, 3–5 January 2007; pp. 621–624.
20. Al-Sarawi, S.F. Low power schmitt trigger circuit. *Electron. Lett.* **2002**, *38*, 1009–1010.
21. Azarmehr, M.; Rashidzadeh, R.; Ahmadi, M. Low-power oscillator for passive radio frequency identification transponders. *IET Circuits Devices Syst.* **2012**, *6*, 79–84.